CORK REGIONAL HOSPITAL
LIBRARY

AF598678

Experientia Supplementum 30

Biochemical Principles of the Use of Xylitol in Medicine and Nutrition with Special Consideration of Dental Aspects

Kauko K. Mäkinen

Institute of Dentistry, University of Turku, Turku, Finland

1978 Birkhäuser Verlag, Basel und Stuttgart

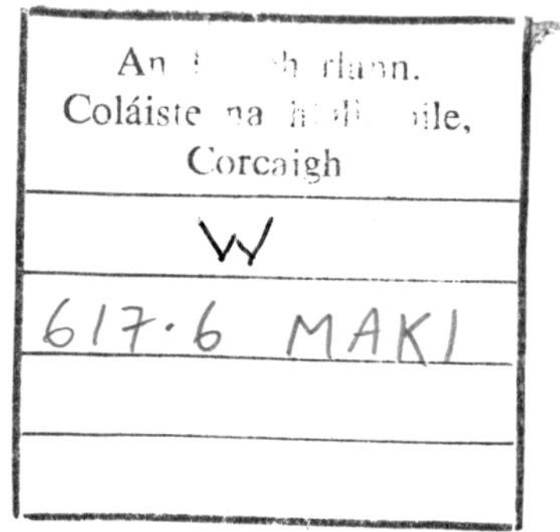

CIP-Kurztitelaufnahme der Deutschen Bibliothek

Mäkinen, Kauko K.
Biochemical principles of the use of xylitol in medicine and nutrition with special consideration of dental aspects. - 1. Aufl. - Basel, Stuttgart: Birkhäuser, 1978.
(Experientia: Suppl.; 30)
ISBN 3-7643-0961-X

Printed in Switzerland

ISBN 3-7643-0961-X

Contents

1. Introduction ... 7
2. Xylitol as a chemical compound ... 9
3. Determination of xylitol ... 15
4. Aspects related to the production of xylitol ... 17
5. Occurrence and significance of polyols ... 19
6. Xylitol in dentistry ... 23
6.1 Basic considerations ... 23
6.11 Microorganisms and various types of caries ... 24
6.12 *Streptococcus mutans* and epidemiology of caries ... 25
6.13 Ecology and metabolism of *Streptococcus mutans* ... 25
6.2 Osmotic pressure of xylitol solutions in caries lesions and oral mucosa ... 28
6.3 Xylitol information of oral bacterium genomes ... 29
6.31 Basic concepts ... 29
6.32 Metabolism of xylitol in oral microorganisms ... 31
6.33 Metabolism of xylitol by other microorganisms ... 34
6.34 Utilization of xylose by microorganisms and animals ... 38
6.35 The possibility of xylitol adaptation in plaque ... 40
6.4 Effect of xylitol on enzymes ... 41
6.41 Inhibition ... 41
6.42 Specificity requirements of enzymes ... 43
6.43 Specifically on sugar transport through bacterial cell membrane ... 45
6.44 Dextran-antidextran system ... 46
6.5 Cariogenicity of polyols ... 46
6.6 Facilitation of mineralization by xylitol ... 48
6.7 Effects of xylitol on plaque and saliva in Turku sugar studies ... 52
6.71 Amino acid composition of oral fluid ... 52
6.72 Invertase-like activity ... 52
6.73 Plaque lactate ... 53
6.74 Oral glycosidases ... 54
6.75 Peroxidase activity ... 54
6.76 Other effects of xylitol ... 57
6.77 Effect of long-term use of xylitol-sweetened chewing gum ... 58
6.8 Conclusions: outlines of the xylitol effects ... 60
7. Xylitol in dietetics and medicine ... 63

7.1 Discovery of enzymes of xylitol metabolism in mammals 63
7.2 Aspects of the metabolism of xylitol in mammals 65
7.21 General features 65
7.22 Glucuronate-xylulose cycle 66
7.23 Assessment of certain metabolic and toxicological studies 68
7.24 Base excess in infusion therapy 72
7.25 Aspects of the 2-year Turku feeding study 73
7.26 Xylitol in erythrocyte metabolism 77
7.27 Effect of xylitol on exocrine gland function 79
7.28 Polyol cataracts 81
7.3 Safety of xylitol 84
7.31 The Australian cases 84
7.32 Other aspects related to the safety of xylitol 91
7.4 Nutritional aspects 94
7.5 Ignored facts 96
8. Conclusion 99
9. Addendum 105
9.1 Physiological and dental aspects 105
9.2 Aspects in parenteral nutrition 109
9.3 Chemical aspects 111
10. List of animal and human studies (Table 8) 113
11. References 145

1. Introduction

Xylitol is a polyalcohol (sugar alcohol) of the pentitol type. Xylitol has been known to chemistry at least from the 1890's. The first preparation of xylitol by sodium amalgam reduction of D-xylose resulted in a syrup-like product [1, 2], but the first successful crystallization, after reduction of highly purified D-xylose, was performed first during the second world war by Wolfrom and Kohn [3]. This product was not, however, a stable form of xylitol. Carson et al. [4] in 1943 were able to obtain stable and crystalline xylitol with a melting point between 93 and 94.5 °C. In spite of its fairly long history in organic chemistry, xylitol can at present be considered a new-comer in medicine and nutrition. Because xylitol is a close relative of sucrose and other dietary carbohydrates, its use in medicine and dietetics is practically the same as for other sugars. Many properties of xylitol, like sweetness, appearance, and physical combustion value are the same as for sucrose.

Because xylitol is a natural product, man has always eaten it daily, although in fairly small quantities. People who habitually consume higher amounts of vegetables and fruits, also obtain xylitol and other sugar alcohols in higher quantities. Livestock and seed-eating birds normally ingest considerable amounts of sugar alcohols, including xylitol. The sweetness of the manna of the Bible was largely based on D-mannitol and most wines contain sorbitol and xylitol. This is to indicate that human and animal evolution embraced various sugar alcohols long ago before any discussions about the properties of these compounds were commenced by man. Virtually all plant material so far studied seems to contain xylitol. According to the present knowledge the richest natural sources seem to be plums *(Prunus domestica)*, strawberries, raspberries, cauliflower and endives, in which the concentration may reach 0.3–1.0 g in 100 g dry material. Considerable amounts of xylitol also occur as an intermediate in the human carbohydrate metabolism, viz. 5–15 g daily.

Until recently the practical importance of xylitol has been minor. This has been due to the small quantities produced and the price of the product. In recent years several human clinical trials have revealed that xylitol may have certain advantageous health effects. Therefore, after the production of small quantities of xylitol in Japan, the Soviet Union, West Germany, and some other European countries, mass production was started in Finland in 1974. This step led to a considerable reduction of the price of xylitol. Simultaneously, opportunities for wider use of xylitol in medicine and nutrition were offered. Because of the rather low amounts of free xylitol in nature, the main present industrial processes start with xylan-rich plant material which is first hydrolyzed to xylose. After reduction of this 'wood sugar', xylitol and other resulting polyols are separated so effectively that the final highly pure and crystalline xylitol contains only trace amounts of mannitol, sorbitol, galactitol, arabitol and possibly some other polyols.

The present treatise was encouraged by the increasing interest in certain proposed extraordinary health effects of xylitol. Such effects may be more easily manifested dentally, but causations between certain medical effects and properties of xylitol have also been revealed. The impressive reduction in the incidence of dental caries in human subjects by a diet in which sucrose was substituted by xylitol or even by very low amounts of xylitol (3–10 g daily) as supplement of a normal diet, seems to be inconceivable in certain circles. However, based on the present knowledge a sufficient explanation to this effect can be provided. Although the dental and so-called medical effects of xylitol are in the present treatise deliberately touched upon separately, one should keep in mind that at the molecular level all classical boundaries between dentistry and medicine vanish. Hence any dentist, physician or dietician should also consider the overall effect of xylitol on human biology.

2. Xylitol as a chemical compound

The crystalline xylitol prepared by Wolfrom and Kohn [3] in 1942 was a hygroscopic and metastable product with a melting point of 61–61.5 °C. The product was formed during prolonged standing of xylitol syrup in an ice box. From this form of xylitol Carson et al. [4] separated approximately 1 year later a stable anhydrous xylitol which was more suitable for chemical and biological studies than the earlier product. The product of Carson et al. [4] was finally recrystallized from ether (methanol and ethanol were also suitable) as colorless crystals with a melting point of 93–94.5 °C. The low-melting monoclinic form usually changes in a few days into the high-melting rhombic form on exposure to the air at room temperature. The crystallization of the low-melting metastable form is easily achieved from alcoholic solution when lath-shaped crystals with oblique ends result. An aqueous viscous syrup also yields complete large crystals on standing several weeks at room temperature.

Stable, high-melting xylitol usually crystallizes as colorless rhombic tablets from methanol, ethanol and aqueous solution. Stable anhydrous xylitol is practically non-hygroscopic. By dehydrating xylitol with sulphuric acid or benzenesulphonic acid, a pure, optically inactive anhydroxylitol can be isolated in crystalline form [5]. This step removes one molecule of water, resulting in a molecule characterized by a DL-1,4 oxygen ring [5].

For identification purposes xylitol can be easily converted to a number of derivatives. Xylitol pentaacetate forms colorless plates with a melting point of 62.5–63.5 °C [4, 6]. Anhydroxylitol was characterized by Carson and Maclay [5] by the properties of three crystalline derivatives, the tribenzoate, tricarbamidate and the monotrityldiacetate. Hewitt and Hudson [7] prepared 2,3,4-triacetyl-1,5-anhydroxylitol, a meso substance, devoid of optical activity. Deacylation gave 1,5-anhydroxylitol which was of meso configuration.

A number of xylitol derivatives has been used in medical experiments. Thiosorbitol and thioxylitol were tested against cadmium poisoning [8]. Several sodium antimonyl polyols were prepared and tested in human filariid infections [9]. Alkaline solutions of polyhydroxy alcohols react with antimony oxide to yield water-soluble complex metal compounds [10, 11]. For example, sorbitol forms a compound of the following type: $C_6H_{13}O_5$–O–Sb–(ONa)–O–$C_6H_{13}O_5$. The metal complex solutions are alkaline (pH approximately 10.5 for NaSb-xylitol in a 0.1% aqueous solution). The complexes are not very stable. Xylitol pentanitrate [2] was shown to be effective against angina pectoris [397].

Xylitol can be condensed with acetone, giving a crystalline, optically inactive diisopropylidene xylitol which after esterification with p-toluensulphonyl chloride in pyridine yields monotosyl diisopropylidene xylitol (unsymmetrical) [12]. Like all acyclic polyols, xylitol has two identical terminal -CH_2OH groups. As the molecule is symmetrical, the use of D or L in the name is unnecessary. The molecular symmetry of xylitol may in certain cases explain its ability to inhibit some enzymes more effectively than other pentitols do. Both ends of the molecule can act in an exactly identical manner thereby significantly increasing the productivity of collisions with specific and unspecific binding sites.

Table 1 gives a list of certain properties of xylitol. Xylitol has a negative heat of solution which is here designated as the difference between the lattice energy and solvation energy. The additional energy, required in bringing crystalline xylitol (rhombic) into solution, is taken from the environment (solvent) and exhibits a cooling effect. This endothermic reaction may have only indirect connections to the dental and medical qualities of xylitol, but it creates advantageous or pleasant organoleptic sensations in persons consuming certain xylitol products. Other sugar alcohols and glucose also display a cooling effect, but clearly to a lesser extent than xylitol.

As in all sugars, the polar groups of xylitol are highly solvated. In an aqueous solution the hydrogen atom of each OH-group is rapidly exchanged with the hydrogen atoms of water. The carbon-bonded hydrogen atoms (C–H) and oxygen atoms (C–O–) are firmly bound. In aldoses and ketoses one oxygen, the hemiacetal oxygen of the carbonyl group is, however, more active than the other oxygens. In xylitol and other related polyols there is no carbonyl group. Sugars in general undergo transformations in water, particularly in the presence of acids and alkalies. These transformations usually involve the carbon of the ketone or aldehyde group. Sugars with no free or reactive carbonyl group can be considered more stable than those bearing a reducing group. Therefore, xylitol is well suited for many food-manufacturing procedures involving heat. For example, it reacts with amino acids only to a low extent when compared to ketoses or pentoses. The fact that xylitol does not enter into Maillard reactions evokes interesting applications in food industry. This is also supported by the heat-stability of xylitol. It is stable, even in solution, at processing temperatures up to 150–180 °C. No hydroxymethylfurfural is formed. Other chemical properties of xylitol have been discussed earlier [13–15].

Alditols may form certain type of complexes with various inorganic polybasic acids (or their salts and anhydrides) in aqueous solutions [16]. Complexes with boric, molybdic, tungstic and other acids, as well as the oxides of antimony and arsenic, have been reported. These complexes are believed to be true esters with one or more molecules of alditols. Consequently, a chelate type of structure may be involved at some point [16]. When such acids or salts are added to solutions of polyhydroxy compounds, the conductivity and acidity of the solution is increased [16, 17]. Simultaneously an increase of the rotation of optically active substances and a marked change of volume result [16, 17]. The salt concentrations required are high from the physiological point of view (for example, 0.1 N ammonium paramolybdate in 0.1 N H_2SO_4). The behaviour of polyhydroxy compounds (like mannitol) in this case is explained on the basis of a tendency for the repulsion of adjacent hydroxyl groups. With open-chain α-glycols the mutual repulsion of the OH-groups with free rotation of the carbon atoms is not said to permit complex formation [16]. Thus very little change in conductivity is noted.

Table 1
Physical, chemical and biological properties of xylitol[1]).

Formula	$C_5H_{12}O_5$		
Molecular weight	152.15		
Appearance	White, crystalline powder		
Crystal system	(i) rhombic, stable (ii) monoclinic, metastable		
Odour	None		
Optical character	Inactive		
Melting range	Rhombic: 93–94.5 °C Monoclinic: 61–61.5 °C		
Boiling point	216 °C (760 mm)		
Solubility in H_2O (g/100 g solution)	4 °C: 55.0 20 °C: 62.8 25 °C: 64.2	40 °C: 74.2 50 °C: 80.0 60 °C: 85.1	
Solubility in H_2O (g/100 g water)	4 °C: 122.0 20 °C: 168.8 25 °C: 179.3	40 °C: 291.3 50 °C: 400.0 60 °C: 571.1	
Solubility in 96% ethanol (g/100 g solution)	25 °C: 1.2		
Solubility in methanol (g/100 g solution)	25 °C: 6.0		
pH in water	5%: 5.6	10%: 5.5	40%: 5.4
Density of solution (weight %)	10%: 1.03 20%: 1.07	40%: 1.15 60%: 1.23	
Viscosity of solution	10%: 1.23 cps 20%: 1.67 cps 40%: 4.18 cps	50%: 8.04 cps 60%: 20.63 cps	
Heat of solution	+34.8 cal/g		
Caloric value	4.06 kcal/g	approx. 17 kJ/g	
Refractive index (weight %)	10%: 1.3471 20%: 1.3620 30%: 1.3779	40%: 1.3951 50%: 1.4132	
Moisture absorption 4 days at RT	60% r.h.: 0.05% H_2O, 90% r.h.: 90% H_2O		
Hygroscopicity	In high relative humidity, xylitol is more hygroscopic than sucrose		
Relative sweetness	Equal to sucrose		

Caramelization	Takes place only if heated several minutes near the boiling point. At about 120 °C no caramelization takes place
Possible impurities	Mannitol, sorbitol, galactitol, arabitol[2])
Fermentation	(i) Rare (microorganisms commensal to man) (ii) More common (among certain soil bacteria and yeasts)
Occurrence in nature	Fruits, vegetables and other plant material; mammalian tissues, microorganisms
Absorption	Virtually all ingested xylitol is absorbed by healthy human subjects at a rate somewhat lower than for glucose
Relation to insulin	Insulin-independent
Metabolic fate	In healthy human subjects most xylitol is oxidized via normal pathways to D-xylulose which is further metabolized to glucose

1) Compiled from Manz, Vanninen and Voirol [14], Kracher [15], Przybilka [31], Beilstein [13] and other literature cited in chapter 2.

2) The presence of impurities depends on raw material and process technique. The named polyols may be present in trace amounts in xylitol made by prehydrolysis of hardwood [33]. The content of xylose is nil in xylitol produced from birchwood in Finland.

When the OH-groups are in a ring compound, α-D-glucopyranose, for example, free rotation of the carbon atoms is not possible and cis-OH-groups now exert a greater tendency than trans to form complexes [16, 17]. The increased conductivity with acyclic polyols (from diols to hexitols) has been explained on the ground of a decreased symmetry with respect to the OH-groups, and, consequently, as a decrease in the repulsive action of the OH-groups. There will be a greater opportunity for complex formation which results in increased conductivity [16, 17]. Sorbitol may form three complex borates. D-Mannitol should form two borate compounds [16]. The ability of xylitol and many other polyols to form complexes with heptamolybdate ions ($Mo_7O_{24}^{6-}$) in 1 M acetate buffer, pH 4.0, also depends on the concentration of phosphate ions. Table 2 shows the relative ability of various polyols to form molybdate complexes.

The ability of polyols to form complexes could be utilized to limit the toxicity of the above-mentioned metal compounds in sudden poisonings, but the pH values of certain parts of the alimentary canal may, however, maintain too weak complexation. The complex formation leads to the practical consequence that, for example, polyols often interfere with chemical phosphorus assays based on the determination of the blue molybdene-containing ions.

The complex formation of polyols may not be reflected in any harmful way in the oral cavity or elsewhere in the human body as a result of xylitol consumption. Molybdene-dependent enzymes are most likely not inhibited in vivo by dietary carbohydrates. The specific resistance of the supernatant fluid of human whole saliva (pH 7.2) of subjects fed on various sugar diets was almost the same, the saliva pools of xylitol-fed subjects displaying only slightly higher resistance than saliva obtained from fructose- or sucrose-consuming persons [91]. Although there are no data supporting the relationship between caries and the ability of sugar alcohols to form complexes, such a phenome-

Table 2
Relative ability of certain polyols to form complexes with $Mo_7O_{24}^{6+}$ ions in the presence of phosphate[1]).

Polyol	Relative complex formation
Galactitol	100
Perseitol	100
D-Mannitol	100
D-Sorbitol	100
D-(+)Arabitol	92
L-(−)Arabitol	91
Xylitol	40
Maltitol	13
Ribitol	12
i-Erythritol	3
Glycerol	0
Ethyleneglycol	0
Water	0

1) Tested at 22 °C in the following reaction mixture (7.0 ml): (a) 1.0 ml of a solution containing 1% ascorbic acid and 7.35 mM KH_2PO_4, (b) 4.0 ml of 4.0 mM polyol (aqueous solution), and (c) 2.0 ml of a solution containing 0.67 M acetate buffer (pH 3.9), 2.7 mM ammonium molybdate (VI) tetrahydrate [$(NH_4)_6Mo_7O_{24}\cdot 4H_2O$] and 0.000167 N H_2SO_4. The reactions were started by adding (c) to a mixture of (a) and (b). The mixtures of (a) and (b) were freshly prepared. Blue color (700 nm) developed at different rates, depending on polyol. The value of 100 was given to compounds which did not increase the absorption in 30 minutes. The value of 0 was given to compounds which had no effect on the development of absorption. Other details in [32].

non may under certain circumstances be important. Polyols may form complexes with Ca^{++}, Zn^{++}, Mg^{++} and other ions important for bacterial metabolism, or polyols may indirectly facilitate remineralization of carious lesions. It is highly likely that polyols, like xylitol, may interfere with the aggregation of plaque bacteria by complexing divalent metal cations in plaque.

Arguments for the structure-supporting properties of polyhydric alcohols (and polyhydroxy compounds in general) have been presented. Two pertinent contributions will be mentioned here. Polyhydric alcohols are relatively strong hydrogen-bonding donors. This characteristic accounts for a certain structure-supporting capability for polypeptides and proteins which contain sites for hydrogen-bond formation [19]. Polyhydroxy alcohols support peptide secondary structures more extensively than halo-alcohols and water [19]. The structures which are supported by ethylene glycol are the right-handed α-helix, β-structure and triple-helical collagen-type conformation. Although the study referred to was performed with ethylene glycol ($CH_2OH\cdot CH_2OH$) as the representative of this family of solvents, the findings cannot totally be ignored from the biological point of view.

Another widely known effect of polyhydroxy compounds (including sucrose) is the capability to protect enzymes against denaturation or other type of inactivation. These effects are most likely related to the colligative properties of the solutions [20]. From the dental point of view it is natural that both cariogenic and low-cariogenic sugars share the above-mentioned protective properties.

As a conclusion, the chemical effects of xylitol in the human body may be attributed at least to the following molecular properties:

- Open-chain structure.
- Absence of reducing carbonyl groups.
- The decisively shorter length of the molecule (approximately 0.29 nm) when compared to hexitols (approximately 0.34 nm).
- The similarities in the configuration at various C-atoms with common sugars (glucose, for example, when open-chain structures are concerned).
- The ability to form complexes with certain metal cations or compounds containing these metal atoms. The complex formation depends upon pH and salt concentrations.

Other consequential qualities of these properties will be listed later. However, the nonexistence of a ring structure, a prerequisite for the action of many enzymes acting on carbohydrates, creates certain hindrances for effective microbiological utilization of xylitol. In cases where the enzymes in question presuppose open-chain structures at certain stages of catalysis, alternative explanations can be provided.

3. Determination of xylitol

In principle, three main basic types of xylitol assay methods have been available. One is based on chromatography, the second on enzyme kinetics and the third is based on titrimetry. Most methods serve the determination of other polyols as well. So, for example, sugar alcohols have been chemically determined by iodometric titration [21, 22] and by gas liquid chromatography after acetylation with acetic anhydride in pyridine [23]. Silylated xylitol has also been analyzed in blood using gas-liquid chromatography [24]. In thin-layer chromatography sugar alcohols have been assayed with a variety of reagents, such as KIO_4-p, p-tetramethyldiaminodiphenylmethane [25], vanillin-perchloric acid solution [26], $NaIO_4$-benzidine reagent [27] and aniline phtalate [23, 538]. The vanillin method was the first which distinguished polyols from ketoses without revealing aldoses. In thin-layer chromatography on Kieselgur G in 0.01 M phosphate buffer, pH 5.0, a recommended procedure involves the development in butanol-acetone-buffer (4:5:1). Xylitol can then be analyzed qualitatively or semiquantitatively by spraying with ethanol-H_2SO_4-anisaldehyde (9:0.5:0.5).

Xylitol has been determined enzymatically with the xylitol-specific enzyme xylitol-NADP dehydrogenase (xylitol:NADP oxidoreductase, EC 1.1.1.10) of guinea-pig liver [18, 28]. The oxidation of xylitol is followed at 366 nm. Another enzymatic method [29] exploits sorbitol dehydrogenase (L-iditol:NAD oxidoreductase, EC 1.1.1.14). The equilibrium catalyzed by the enzyme favours the formation of xylitol, but by adding hydrazine sulphate it is possible to trap D-xylulose as its hydrazone. Thus a quantitative turnover of xylitol results. As the enzyme involved is not very specific, the procedure involves the elimination of fructose with hydrazine prior to the use of dehydrogenase. Other enzymic methods which use sorbitol dehydrogenase have also been published [30]. Specific identification and assay methods were reviewed by Przybilka and Linke [31].

There is a need for a rapid and simple colorimetric assay for xylitol and other polyols. The chemical properties of these compounds allow an additional possibility for their determination. Because alditols form complexes with various inorganic polybasic acids, their salts and anhydrides [16, 17], a colorimetric determination of xylitol and other sugar alcohols as their molybdenum complexes, for example, would thus be possible. A satisfactory colorimetric method has been developed [32]. Most of the listed methods suffer from unspecificity. The use of specific xylitol-oxidizing enzymes in addition to gas chromatography, appears to be convenient methods. Histochemical methods for the determination of polyoldehydrogenase with xylitol have been studied [398, 399].

Xylitol can be extracted from most foodstuffs with a mixture of acetone and methanol (1:1), containing 2–3% water. Solid food should be finely ground and the resulting powder can be dehydrated with benzene in a rotatory evaporator. Liquid samples can be freeze-dried before extraction.

4. Aspects related to the production of xylitol

The history of xylitol was fairly eventless until its potentialities in medicine and dentistry emerged. Production procedures were developed and improved and up to the 1970's the patent literature had already been enriched by several applications [33]. The production of xylitol has been based mainly on hydrogenation of D-xylose obtained from prehydrolysis of various xylan-containing plant materials (birchwood, cottonseed hulls, coconut shells, etc.). Other xylan-containing sources are straw, pecan and almond shells, corn stalks, beechwood, etc. In general, raw material for xylitol production is abundant. Thus, production of xylitol from waste materials might help to solve some pollution problems. At present the bulk of pure xylitol available is made in Finland from birchwood chips [33]. A key step in the process is the separation of polyol mixtures by ion exchange chromatography [33].

A large variety of polyalcohols are also produced in rather good yields by aerobic dissimilation of various pentoses and hexoses by yeast cells [34–46]. Microbial production of xylitol from glucose via D-arabitol and D-xylulose is also possible using a sequential fermentation process involving three types of microorganisms in subsequent steps [47]. The yield can reach almost 12%. In the biological production of xylitol a particular problem may arise from the presence of harmful bacterial components in the final preparations. Ion exchange chromatography serves in this case as an advantageous purification procedure. Certain other [48, 49] microbiological aspects, related to microbial formation of xylitol, can also be mentioned. Although Scharkow [50] presented in 1963 how it is possible by using selected plant material and conditions of hydrolysis, to obtain various monosaccharides and to convert them to polyols, similar technical and chemical development had simultaneously and previously taken place within carbohydrate industry in various countries. Knowledge of certain important key steps has been in the possession of xylitol-producing enterprises. The synthesis of xylitol from hydrocarbons of oil and related sources should be possible. The chemical steps are known, but the procedure involved may prove to be expensive.

In order to be able to explain the advantageous dental and medical effects of xylitol, it is essential to emphasize that the product used in recent human clinical trials in Turku did not contain to any significant extent fluorides, thiocyanate or other inorganic ions, important in oral biology [90–94, 99–101]. The separation of xylose from xylitol is easily accomplished in mass production. The xylitol prepared from birch, for example, meets the strictest criteria of purity.

5. Occurrence and significance of polyols

Not too many years ago authoritative text books in biochemistry indicated that xylitol does not occur in nature. As a result of increased interest in sugar alcohols and improved assay methods, it is now realized that xylitol is a ubiquitous carbohydrate. The present literature on the occurrence of xylitol in nature will most likely increase rapidly during the forthcoming years. Xylitol and certain other polyols are of vital importance to the life of numerous animal, plant and bacterial species. As a class, the polyols appear to be capable of functioning as nutritive substrates for a large variety of microorganisms. However, no single organism seems to be capable of utilizing every polyol. In the higher plants and particularly in the fruits, polyols appear to function as reserve carbohydrates, the quantities being seasonal and often becoming less as the other sugars increase during the ripening process.

Xylitol was perhaps first detected in certain lower plants, such as lichens, seaweed and yeast [51]. It was then discovered in champignons *(Psalliota campestris)* [28] and other mushrooms [27] in amounts of 100–130 mg per 100 g dry weight. In extensive studies of Washüttl et al. it was shown that xylitol occurs in most plant materials studied [27]. The highest analyzed quantity was 935 mg xylitol per 100 g dry weight, which was discovered in plums. It is interesting to note that galactitol was also detected in fairly high amounts in yoghurt, artificial honey, plum jam, fig coffee (surrogate prepared of figs), juniper berry electuary, baker's yeast, etc. [27]. The presence of xylitol in fruits leads to the fact that man also eats xylitol daily in products made of fruits, such as certain wines, jams, marmalades, squashes, etc. It is expected that certain variations will in the future be detected in the content of xylitol in plants, depending on geographic location, growth conditions, and other chemical and physical determinants. Table 3 gives the xylitol content of some fruits, vegetables and foods.

In some cases special functions of polyols in lower organisms have been suggested. In *Aspergillus clavatus* the main function of mannitol is that of a storage compound, possibly connected with conidiation, while ribitol appears to be primarily involved in hydrogen-acceptor mechanisms [52]. The enzymology and production of polyols in fungi has been the object of many studies [45, 54, 55]. Mannitol has been implicated in sporulation of higher fungi [56] and as a carbon source in germinating *Aspergillus oryzae* conidia [57]. Mannitol and other polyols in microorganisms were considered either to be involved in

Table 3
Occurrence of xylitol in some fruits, vegetables and related products. Expressed as mg per 100 g of dry matter, or as mg per 100 g wine or juice[1]).

Plums (*Prunus domestica* ssp. *italica*)	935
Strawberries (*Fragaria* var.)	362
Cauliflower (*Brassica oleracea* L. var. *botrytis*)	300
Lamb's lettuce (*Valerianella olitoria* L.)	273
Raspberries (*Rubus idaeus* L.)	268
Endives (*Cichorium endivia* L.)	258
Egg plant (*Solanum melongena* L.)	180
Lettuce (*Lactuca sativa*)	131
White mushrooms (*Boletus edulis* Bull.)	128
Apple wine	120
Spinach (*Spinacea oleracea* L.)	107
Pumpkins (*Cucurbita pepo* L.)	96.5
Kohlrabi (*Brassica oleracea* L. var. *gongylodes* L.)	94
Fennel (*Foeniculum vulgare* Mil.)	92
Onions (*Allium cepa* L.)	89
Carrots, fresh (*Daucus carota,* L.)	86.5
Red cherry jam	56
Morello cherry jam	54.5
Leeks (*Allium porrum* L.)	53
Black currant jam	34
Bananas (*Musa sapientium* L.)	21
Canned pineapple *(Ananas sativus)*	21
Chestnuts edible *(Castanea vesca)*	14
Carrot juice	12
Brewers' yeast	4.5
Liquorice	4.5
Cornmeal	0.5

1) From Washüttl et al. [23, 27].

hydrogen-acceptor mechanism [58] or to act as reserve carbohydrates [54, 59]. Certain polyols (at least glycerol and ribitol) are components of teichoic acids which are polyolphosphate polymers found in bacteria, usually in close association with the cell walls.

A number of lower organisms known to be involved in xylitol metabolism or catabolism have also been mentioned in the literature citations in the section 'Aspects related to the production of xylitol'. Accordingly, many lower forms of life are also genetically informed about xylitol. Other organisms, involved in the production of xylitol, are, for example, the yeast *Pichia quercibus* (Phaff et Knapp), isolated from slime flux of black oak [49], and *Penicillium chrysogenum* [48]. Both form xylitol from D-xylose. Onishi and Suzuki studied 128 yeasts for their ability to produce xylitol directly from glucose [47]. Arabitol was the only pentitol produced, but neither xylitol nor ribitol was found. Screening for yeasts producing xylitol from D-xylulose showed, however, that 27 types of yeasts were able to form xylitol. Certain yeasts produced both xylitol and D-arabitol [47]. These findings indicate that xylitol may occur rather widely spread (but in very low concentrations) in earth and microorganisms, because of the abundance of a suitable starting material (xylan and/or xylose) in their environment.

In spite of the fact that xylitol is produced from precursors by so many yeast and certain other microorganism species, the organisms commensal to man appear to attack xylitol only to a very restricted extent. This also stands for many microorganisms which are harmful in food processing and storage.

Xylitol occurs as a natural intermediate in the carbohydrate metabolism of man and other mammals. Later studies will show its contribution to the carbohydrate metabolism of other species as well. The physiological transformation of xylitol through the glucuronic acid cycle [257, 258] amounts in healthy subjects to 3–10 g/day [608, 609], but taken as a whole, the endogenous transformation in man is 5–15 g/day. At any given moment, human blood seems to contain xylitol 0.3–0.6 mg per 1 [444]. After oral administration of larger amounts of xylitol, the concentration may not exceed 50 mg/l. However, in the portal vein the concentration may be temporarily higher.

6. Xylitol in dentistry

According to the present understanding, ingestion of xylitol has an inhibitory effect on cariogenesis in man. This effect can be explained in physicochemical, biochemical and microbiological terms. In this chapter the biological effects of xylitol in human oral cavity (and elsewhere in the body, when applicable) are classified into two groups:

I. Primary effects and properties. These are here designated as the immediate and initial reactions at the molecular level, caused by the molecular properties of xylitol and by its close proximity to oral tissues and constituents (or other constituents in the body) during and after intake of xylitol.

II. Secondary effects and properties. The secondary effects are natural consequences of the primary ones, i.e. of the molecular properties of xylitol. These effects include most secondary microbiological, enzymic and chemical changes in whole mouth saliva and dental plaque, gingival exudate, as well as in the intestines. The secondary effects also comprise all consequential phenomena in the liver, salivary glands and other organs which are dealing with the metabolic products of xylitol.

The above classification is to a certain extent artificial and too strict and in many cases it will be difficult to show which category is involved. Consequently, the classification merely attempts to arrange the subjects in logical order and to emphasize the molecular differences between pentitols and hexitols and those between sugar alcohols and other carbohydrates. To illustrate this, the *appearance* of an enzyme or protein in elevated amounts in saliva is regarded as a consequential, secondary reaction. This is naturally preceded, however, by certain primary reactions in the gastrointestinal tract, including salivary glands, leading to the appearance of elevated enzyme activity. It is evident that we know at present more about the manifestations of the presence of xylitol in diet than about the underlying molecular mechanisms, or primary effects.

6.1 Basic considerations

A number of chemical properties required in the explanation of the dental effects of xylitol were mentioned in chapter 2. It should particularly be emphasized, from the dental point of view, that the difference between the length of the pentitol and hexitol molecules has to be regarded as one decisive factor. The subsequent chapters describe a number of phenomena of quite varying nature. In spite of the apparent inconsistence involved, the factors which will be touched upon, are, however, dentally interrelated.

For a full comprehension of the effects of xylitol in dentistry, it is necessary to review briefly some pertinent concepts about the cariogenicity of certain oral microorganisms. For this purpose the aspects shown below will be catalogically touched upon (additional information has been provided in two recent reviews [61, 62]):

6.11 Microorganisms and various types of caries
6.12 *Streptococcus mutans* and epidemiology of caries
6.13 Ecology and metabolism of *Streptococcus mutans*

6.11 Microorganisms and various types of caries

(1) Dental caries refers to a destructive process affecting the teeth. It is to a certain extent similar to other infectious processes which occur elsewhere in the human body. An infectious disease may frequently be caused by a fairly specific etiologic agent, but in certain instances this is not the case.

(2) In dental caries at least one very potent pathogen is involved, *Streptococcus mutans*. It may particularly be involved in smooth-surface caries [63], contributing to the plaque growth on these areas of teeth. Usually, microorganisms promoting plaque growth have been considered causative factors in dental caries. However, other microorganisms can also become involved depending on circumstances. Similarly, pneumococcus is not the only etiologic agent of pneumonia. The relationship between meningitis and meningococcus is also similar. These facts do not reduce the significance of pneumococcus and meningococcus in these diseases. The predominant organism in the throat during diphtheria is said to be streptococcal, whereas the causative agent, *Corynebacterium diphtheriae*, is present in relatively small numbers [63]. Thus the mere presence of streptococci in large numbers does not necessarily make these bacteria the main or only causative agent in dental caries. The idea presented by Kleinberg [613] is particularly attractive in this context. According to him dental caries is caused not so much by excessive plaque glycolysis, but rather by an inability of the plaque bacteria to form enough base from salivary substrates to neutralize the acid that is formed during degradation of dietary sugars. The consumption of a xylitol diet was associated with an increase in the concentration of basic amino acids in whole saliva [91].

(3) Gnotobiotic studies have clearly indicated that certain strains of lactobacilli, for example, *L. casei* and *L. acidophilus*, can induce fissure caries in gnotobiotic rats. It has been suggested that lactobacilli may be involved in extension of existing caries lesions [64] and in fissure caries [63].

(4) Experiments with laboratory animals and humans indicate that a number of different bacterial types have the potential to destroy cementum and penetrate the depths in dentine. Such organisms have been primarily filamentous *Actinomyces viscosus* and *A. naeslundii.* Keyes [65] has reported about senile root caries among a Papuan population, which was ascribed to the action of *Actinomyces.* It is thus wrong to overemphasize the role of *Str. mutans.*

(5) Non-plaque-forming streptococci may also be involved in dental caries. For example, a few strains of the dextran-forming *Str. sanguis* may initiate minor decay.

6.12 *Streptococcus mutans* and epidemiology of caries

(1) There is very strong evidence that at least *Str. mutans* is linked to the initiation and development of dental caries, although its role has been greatly exaggerated.

(2) Although *Str. mutans* seems to preferentially colonize tooth surfaces, it can also be found in the intestinal tract of individuals whose dental plaque harbors this organism [66].

(3) *Str. mutans* has been found wherever it has been looked for throughout the world [67]. It is most likely not associated with one's standard of living, because the organism has been found in most affluent societies [68–70] as well as in primitive peoples of New Guinea and isolated villages in South America [71, 72]. The organism is most likely endemic in the world. Thus it is not introduced into a society by sucrose-eating civilized man [67, 73].

(4) In almost all cases the organisms can be found in carious lesions [74, 75]. However, the presence of this organism in the human oral cavity and on a particular site of tooth surface does not necessarily mean that caries will with certainty develop at that site [67]. Other determining factors are the state of tooth maturation, host resistance, cariogenicity of the diet, concentration of fluoride and other trace elements in drinking water, etc.

6.13 Ecology and metabolism of *Streptococcus mutans*

(1) The ability of *Str. mutans* to initiate dental caries is most likely related to its ability to colonize and accumulate on the tooth surfaces, and to its ability to contribute to dental plaque formation. Sucrose is required for the colonization and accumulation.

(2) *Str. mutans* is able to synthesize extracellular polysaccharides specificially from sucrose, but usually not from other sugars. With the aid of these sticky and adhesive polysaccharides *Str. mutans* grows as a clammy mass attached to the tooth.

(3) *Str. mutans* synthesizes two general types of polysaccharides from

sucrose: glucans and fructans. Both are most likely involved in demineralization. Some glucans contain more α-1,3 links than α-1,6 links. Such polymers have been termed mutans. They are more insoluble than normal dextrans. Particularly glucans have been claimed to be decisive factors in the adhesive interactions of *Str. mutans*. Xylitol reduces this adhesiveness by increasing the proportion of soluble polysaccharides in plaque.

(4) In addition to sucrose being required for plaque formation by *Str. mutans,* sucrose has also been found to enhance infection by this organism [76]. *Str. mutans* grows almost equally well in a glucose broth as it does in a sucrose broth. However, it can not readily colonize humans consuming a diet rich in glucose. The requirement of sucrose for the adhesive interactions of the organism indicates that the ability of the cells to attach to a surface is an important ecological determinant [76].

(5) *Str. mutans* does not readily colonize the mouth under normal conditions. Cells transmitted into the mouth are usually readily cleared [76]. Consequently, the organisms are not easily transmitted from one individual to another. Transmission occurs, but it seems to be infrequent [76]. Implantation is, however, improved by sucrose.

(6) The organism colonizes in a localized and consistent manner. This may partly explain its high cariogenic potential. Colonies of *Str. mutans,* embedded in adhesive and protective carbohydrate polymers, are located on the tooth surface, rather than at the periphery of plaque. Thus tongue and cheek movements, oral hygiene and other mechanisms are not able to remove these colonies readily. Acid production within the gel mass may thus lead to the development of a carious lesion which is first detected as a white chalky spot, characterized by a highly local demineralization.

(7) Because the organism colonizes in a localized manner, and is not quickly transmitted from one tooth surface to another [76], one could temporarily 'sterilize' a tooth surface which could then remain free of the organism for certain significant periods of time. Gibbons [76] has found that the use of disinfecting agents appears to abolish *Str. mutans* for months. It is conceivable that when the growth of *Str. mutans* is reduced by xylitol, there would be a similar, although not necessarily an equally long-term effect. Consequently, in a figurative sense, xylitol may temporarily 'sterilize' a tooth site.

(8) When the plaque has been firmly established with the aid of sucrose, *Str. mutans,* particularly in occlusal pits and fissures and other retentive areas, may initiate dental caries even in the absence of sucrose, by using glucose and other easily fermentable carbohydrates. However, on smooth surfaces the accumulation seems to require sucrose.

(9) The enzymes involved in the synthesis of glucans have been termed dextransucrases. They most likely act as glucosyltransferases (EC

2.4.1.5; α-1,6-glucan: D-fructose 2-glucosyltransferase). Their activity requires a glucan acceptor. Free fructose is left for dissimilation. There are multiple glucosyltransferases produced by *Str. mutans* strains [77, 78]. The genus *Leuconostoc* is another producer of dextransucrases.

(10) *Str. mutans* is a homofermentative cell [79]. Fermentation of one mole of glucose produces almost two moles of lactate [80–82]. Incubation of *Str. mutans* with low concentrations of sucrose results in the conversation of almost all of it into lactate (four moles of lactate per one mole of sucrose) [83].

(11) In addition to glucosyltransferases, invertase and sucrose permease activity are also involved in the utilization of sucrose by the cells. It is probable that both of these enzymes are inducible [84–87]. The permease-invertase system brings sucrose from the outside to the inside of the cells. Sucrose is converted to equimolar amounts of glucose and fructose [79]. The importance of bacterial invertase-like enzymes in caries has been indicated in several connections [61, 62, 88–94]. The consumption of a xylitol diet lessens the activity of these enzymes in plaque and whole saliva [88–94]. *Str. mutans* does not seem to have a xylitol permease.

(12) In addition to extracellular polysaccharides, intracellular glycogen-like storage polymers also contribute to cariogenicity. Both fructan and intracellular polysaccharides may serve as substrates for prolonged acid formation in the absence of exogenous sucrose or other fermentable dietary sugars. Thus these polymers may also play a role in demineralization.

(13) A nutritional characteristic which serves to distinguish cariogenic *Str. mutans* strains from other homolactic oral streptococci, is the ability of the former to use mannitol and sorbitol as a primary source of energy [77, 95–97]. The ability of the cells to grow on mannitol or sorbitol may not contribute to the caries-inducive potential of these organisms, because spontaneous mutants of *Str. mutans,* unable to ferment sorbitol and mannitol have not lost their ability to cause sucrose-dependent caries in animals [95, 98]. However, fermentation of these hexitols may contribute to the survival potential of *Str. mutans* in the oral environment [95]. Sorbitol may, however, decrease the pH value of a plaque suspension even to 3.9 [614].

(14) The anaerobic fermentation of mannitol and sorbitol by *Str. mutans* results in an additional mole of NADH (when compared to the anaerobic catabolism of sucrose and glucose by these cells). The additional NADH is generated at the level of an inducible hexitol phosphate dehydrogenase. The additional NADH in *Str. mutans* is reoxidized by an inducible NAD-linked alcohol dehydrogenase [95].

The above brief list indicates that partial replacement of dietary sucrose with certain natural and physiological sweeteners which are not fermented, might prove effective against the development of caries.

Before these considerations are more closely examined in light of certain xylitol effects, another aspect of physicochemical nature will be dealt with in the subsequent section. The recent [99–101] findings that the consumption of surprisingly low amounts of xylitol practically eradicates dental caries [102], prompt at first a wider discussion about the properties of xylitol.

6.2 Osmotic pressure of xylitol solutions in caries lesions and oral mucosa

The osmotic pressure of sugars and other solutes increases with increasing temperature and concentration of the compound (up to a certain limit). The osmotic pressure of a sugar solution is not dependent on the solvent (saliva). The osmotic pressure (π) can be given as $\pi = nRT/V$ where n is the number of moles (g/M), R is the gas constant, T is the absolute temperature, and V is the volume. At fixed concentrations the lower molecular weight of xylitol (152) leads to a higher osmotic pressure than obtained with sorbitol (182) or sucrose (342).

The osmotic pressure follows the above van't Hoff's equation only in very dilute solution. At higher concentrations the Morse equation and others should be considered (for example, those taking into account the vapour pressure)[1]). It should also be noticed that osmotic pressure only occurs when there is a semipermeable membrane between two solutions of different concentrations (or between the solution and the solvent). On that side of the membrane where the concentration of the solvent is higher there should be more collisions against the membrane, which results in the passage of the solvent to the compartment where its concentration is lower. According to another theory a liquid solvent is not able to penetrate through semipermeable membranes, but vapour molecules can (the solvent is distilled from the pure phase to the solution).

The structures involved in the carious dentine and carious enamel are covered by a hydration layer and adsorbed anions and cations, as well as by pellicle material and other organic integuments through (and in) which the sugars should show their effects. As small molecules, mono- and disaccharides are able to diffuse through these films, provided that they are not all immediately used by microorganisms. The capillary pressure maintained in these structures, providing possibilities for sugar translocation, should now be an interesting factor. The capillary pressure in pores and capillary-like structures present in outer enamel, and enamel and dentine caries lesions, may under given conditions overcome the hydraulic pressure in the sugar solutions of certain concentrations. Present information indicates that the consumption of a xylitol diet induces rehardening of carious lesions. In case a dietary carbohy-

1) For the present purpose the differences between the values of π calculated with various equations for a sugar solution at a concentration appearing in saliva, can be ignored. For example, for a 1 M sucrose solution (25%, w/w), the measured π at 30 °C is close to 27 atm. The calculation on the basis of various equations gives values between approximately 22 and 27 atm. For more diluted solutions the deviations decrease.

drate is not able to lead to a local decrease of pH in a lesion or on a tooth site, and if it is not readily attacked by cariogenic bacteria, the difference in the osmotic pressure between various sugar solutions should also be taken into account when explaining the remineralization effects of xylitol. If the penetrating sugar is nonfermentable and cariostatic (an opposite to sucrose), its presence in carious lesions may be advantageous from the dental point of view. Liquids have been found to penetrate into the surface of carious lesions by capillary actions [615]. Effects of capillary penetration at sites of caries susceptibility have been reviewed [616].

Thin, membrane-like structures (although deviating from ideal semipermeable model membranes), lining microcrevices and pore-like microscopical sites, may then participate in the translocation of calcium and phosphate from saliva to carious enamel and dentine lesions. The increase of the concentration of calcium and phosphate in saliva, and the maintenance of the pH of plaque interface at a higher level on xylitol consumption, most likely lead to an increased transfer of calcium and phosphate ions from saliva to the solid phase. Facilitation of mineralization related to xylitol is dealt with later.

Another aspect of somewhat different nature can be added. Both keratinized and non-keratinized oral mucosa may act as a membrane system with an outer permeable and an inner impermeable phase. The outer phase is said to consist of a hydrated surface layer which shows a passive transport to water, and Na^+ and K^+ ions [103]. The inner phase can be recognized after a partial dehydration of the surface. If the pressure at transport exceeds the mechanical resistance of the mucosa, the inner phase shows only a transport of water. Although the mechanism may not be this simple, it may be worth considering in light of the proposed keratinization-promoting effects of xylitol consumption. Less inflammatory changes in the gingivae and elsewhere in the oral cavity appear as a result of xylitol consumption in man compared to sucrose consumption. The above-mentioned outer permeable phase should react in different ways to the presence of sugars of different molecular weight, if the sugar is consumed in higher quantities during a longer period. It would be of particular interest to elucidate the possible relationship between the consumption of xylitol and the secretion of the epidermal growth factor, stored in the convoluted tubules of the salivary glands.

6.3 **Xylitol information of oral bacterium genomes**

6.31 Basic concepts

Let us consider a microorganism cariogenic to man. As long as this type of a host-parasite relationship has existed in the present form, the cariogenic organisms have transmitted information from generation to generation so that only negligible alteration of the final structure of the cells and enzymes has taken place. The molecules that transport cariogenicity in microbes are particular DNA fragments which control the biosynthesis of the cytoplasm, capsular or envelope proteins responsible for cariogenicity. Because the cariogenic microorganisms are unicellular, it is necessary to consider the following phenomena:

(1) The problem of adaptation to the physiological environment (oral cavity), i.e. the response of the cariogenic organism to changes in the growth medium. A change significant enough would be a constant exposure of the organism to higher concentrations of xylitol, or an increased frequency of intake of xylitol.

(2) The problem of the changes in the selective permeability of the cariogenic microorganisms to extracellular sugars, like xylitol.

When considering these two problems, it is necessary to recall a few fundamental properties of the mechanisms involved in the transmission of genetic information [104–108]. There are two fundamental concepts shown below:

(1) One-gene-one-enzyme (or one-gene-one-polypeptide chain) hypothesis which claims that each gene is responsible for the synthesis of a specific polypeptide. The genes work by controlling the sequence of amino acids in proteins. A gene and its polypeptide product are colinear, i.e. mutations that map at an end of a gene affect the amino acid sequence at an end of the polypeptide (which may be an enzyme, like glucosyltransferase, involved in cariogenesis).

(2) Genes controlling a series of related biochemical reactions frequently map adjacent to each other.

Considering the use of xylitol (or of any other sugar) by bacteria, or suitability of the sugar as an extracellular nutrient, it may be required that adjacent genes control the synthesis of enzymes responsible for a biochemical sequence of reactions involved in xylitol utilization. Thus the genes related to cariogenicity of an organism may map adjacent to each other, for example, $A \rightarrow B \rightarrow C \rightarrow D$. Here each arrow stands for an enzyme involved in cariogenicity. A, B, C and D stand for intermediates involved in the pathway. The biosynthesis of a group of enzymes required in the sequential metabolism of xylitol (or any other sugar) is turned on or off in a coordinated manner. This means that *induction* and *repression* are involved in a coordinated way in responses of the cells to chemical changes in the oral cavity (coordinated induction and repression).

The chromosome of a cariogenic microorganism is organized in operons. Operons are genetic units consisting of adjacent genes which function coordinately under the joint control of an operator and a repressor. An operator is a region in the chromosome, which most likely interacts directly with a specific repressor, controlling in this way the action of an adjacent operon. The final result is the synthesis of specific enzymes directly or indirectly involved in cariogenicity. More than ten operons have been studied in some detail in certain microorganisms (*E. coli* and *Salmonella*). Of these operons several are concerned with the metabolism of sugars. For example, one operon is responsible for lactose metabolism (Lac region), one for galactose metabolism (Gal region), one for arabinose (Ara region), etc. These operons are inducible by their substrates or their structural analogues. Therefore, in case specific operons for the utilization of xylitol in a bacterial chromosome exist, the Xyl regions should also be inducible by xylitol.

Inducibility and repressibility in a cariogenic microorganism are controlled genetically. The genes in question are regulator genes. They map in the chromosome in a position not necessarily linked to the operon comprising the structural gene affected. The regulator gene controls the synthesis of a product (the aporepressor) which in turn interacts with the operator. In the case of induction this interaction is hampered and in the case of repression it is facilitated. This is caused by the simultaneous presence of a regulatory compound. Consequently, induction and derepression are said to be formally and mechanistically equivalent. Because induction and repression of enzymes in microorganisms are very common, one is forced to assume that cariogenic and other oral bacteria undergo such phenomena exceedingly frequently. It is possible, however, that even during a longer period only a smaller number of these are revealed due to the insignificant bearing of most of them in oral biology.

6.32 Metabolism of xylitol in oral microorganisms

The manifestation of adaptation to the changes in physiological environment (oral cavity) involving enzyme induction would thus require the existence of a specific Xyl region, or an unspecific polyol region, which would be directed to the utilization of xylitol. Xylitol would also be replaceable with other polyols. In these cases the process would follow the usual steps as involved at the molecular level in induction by other sugars and other microorganisms. The question is now whether cariogenic and other oral microorganisms possess an appropriate inducible or derepressible gene region for the synthesis of enzymes needed in the exploitation of xylitol. All available information about the acid production from xylitol, and about fermentation and use of xylitol in the human oral cavity strongly indicates very low utilization of xylitol in this respect [539, 570].

Consequently, when the metabolism of xylitol by human oral bacteria is insignificant, the literature dealing with this is understandably scant. In vitro studies on *Str. mutans* (Ingbritt) showed that xylitol inhibits the growth of the cells [109] and causes a delay in the attainment of the stationary growth phase and in the formation of aldolase when compared to conditions with no xylitol added. Simultaneously, an induction-like phenomenon of aminopeptidase-like enzymes was revealed [109]. However, after 4–5 months' continuous presence of xylitol in the cultures, normal growth and acid production were induced.

The above study led the author to conclude that a 4–5 months' continuous presence of xylitol in the stab cultures might have led to an adaptation. A series of subsequent experiments showed however, that no adaptation of the cells to use xylitol had taken place. The cells of *Str. mutans* were shown to grow identically in the presence of xylitol as the only added carbohydrate, and without any added carbohydrate at all [110]. Universally labelled ^{14}C-xylitol was not incorporated into the cells to any measurable extent. The true adaptation involved synthesis of extracellular proteinases (or, at this stage of the study, unspecific hydrolases) which obviously used proteins or peptides of the medium as substrates for energy-requiring reactions of the cells. This lowered the external pH and explained the results previously obtained [109]. The cells were shown to suffer more from the absence of glucose than from the presence of xylitol [111]. Glucose-deprived cells thus behaved as true proteinase-producing ones, xylitol acting as an inert, practically innocuous compound in the medium [111]. Related results were obtained with certain strains of oral *Candida* [112]. The growth of the cells in xylitol alone was slow and the maximum cell yield after 72-hr growth decreased with increasing concentrations of xylitol. Continuous 4.5-year use of xylitol did not reveal any bacterial adaptation in human dental plaque [614].

The studies [110, 111] which described the xylitol-dependent increase of the extracellular proteinase (or unspecific hydrolase) activity involved the maintenance of the cells of *Str. mutans* for 18 months at +4 °C with monthly transfers. The cells were continuously kept in the presence of 0.25% xylitol. No evidence of adaptation of the cells to metabolize xylitol was found. Instead, the extracellular proteolytic activity against several proteinase substrates was noticeably increased in the xylitol media. This phenomenon could be termed induction. These suggestions can be criticized as to the following points:

- Can one expect any adaptation at +4 °C?
- Because xylitol is completely ignored by the bacterium, it may be erroneous to conclude that xylitol increases the proteinase production.

It is obvious that the growth of *Str. mutans* at +4 °C is extremely low. However, this temperature does not make the cell wall impermeable or inactive with regard to possible slow translocation of sugars. As to the second point, the increase in the proteinase activity could be explained as catabolite repression by glucose. It is also possible that the strong acid formation from glucose may destroy a part of the proteolytic enzymes, while the xylitol media in which peptides and proteins are metabolized, may become alkaline. This would mean that xylitol has no direct effect on proteinase production. The indirect effect is, however, very pronounced, as shown in figure 1. The acid-lability of the enzymes hardly explains the results. It is more likely that the question was about the search for nutrients.

Recent studies with isotopically labelled compounds indicate that human dental plaque does not seem to possess specific binding sites for xylitol to any significant extent when compared to other common carbohydrates (sucrose, glucose, fructose, sorbitol, mannitol) [113, 539]. This may be regarded as an important property, because the translocation of sugars through bacterial cell wall generally involves binding of the nutrient to specific sites in the wall. The failure of xylitol to bind itself and to undergo degradation in

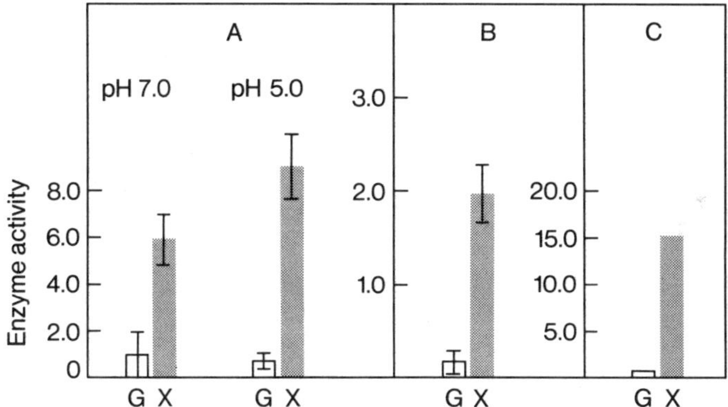

Figure 1
Proteolytic activity of *Streptococcus mutans*. The rate of the hydrolysis of casein (A) at pH 7.0 and 5.0, of a chromophore-collagenase substrate (4-phenylazobenzyloxycarbonyl-L-prolyl-L-leucylglycyl-L-prolyl-D-arginine dihydrate) (B) and of native collagen (C), both at pH 7.0, by enzyme preparations of the extracellular phase of cultures containing either glucose (G) or xylitol (X). The casein-splitting activity is expressed as units/(mg protein × mg dry weight). The hydrolysis of the synthetic substrate was calculated as μmoles/(min × mg protein) and that of the native collagen as ng of liberated hydroxyproline/(min × mg protein), given per mg dry weight of the cells [227]. The enzymes were not denatured by the low pH values of the glucose media.

oral bacteria explains why Hassell [116] was not able to show any critical pH-reduction after rinses with 10% xylitol solutions.

The recent Turku sugar studies showed that the consumption of a xylitol diet reduced the incidence of oral *Candida*, streptococci and acid-forming bacteria [114, 115]. No evidence was obtained for the involvement of oral microorganisms capable of decomposing xylitol. In a few cases very low acid production from xylitol was encountered in freeze-dried plaque samples of the Turku sugar study. The subjects involved did not belong to xylitol, but to sucrose and fructose groups. Animal and microbiological experiments support these findings [117–119]. The Turku sugar study results were also in accordance with those of Karle and Gehring [117, 118] who showed that plaque samples of 'xylitol rats' did not contain *Str. mutans*. 10 or 20% xylitol in diet strongly reduced streptococci in microbiological samples of rat molars [548].

The above findings were also supported by experiments of Gülzov [120] who showed with the Warburg technique that, compared with sucrose, the decomposition of sorbitol by human saliva was somewhat delayed, but continuous and constant. On the other hand, xylitol did not show any conversion. Gülzov reinforced these results later [121].

The activity of xylitol dehydrogenase in human dental plaque and whole saliva is very low or practically nil [89, 91]. The activity of sorbitol and mannitol dehydrogenases is clearly higher. In general, the metabolism of these two latter polyols by dental plaque and oral bacteria is better known than that of xylitol. Plaque ferments hexitols mostly by glycolysis with the production of organic acids, ethanol and other compounds [122–125]. Hexitols are phosphorylated before oxidation. The hexitol phosphate dehydrogenases have been found to be inducible [125].

Polyol dehydrogenases may make an important contribution to the maintenance of the redox potential in mammalian, insect and bacterial cells [400]. The significance of these enzymes in human oral cavity is, however, normally small. The importance may increase on continuous and progressive intake of hexitols and pentitols. 4.5-year continuous use of xylitol did not, however, show any increase in plaque xylitol dehydrogenase activity [614].

A practical and advantageous consequence of the virtually inert properties of xylitol in the human oral cavity is the 30–55% reduction in plaque fresh weight, reported in several separate studies [90–94, 126, 127]. This reduction was obtained although the subjects refrained from all oral hygiene procedures. The mere chewing of a xylitol-containing chewing gum [92–94] or rinsing the mouth with a 10% aqueous xylitol solution five times a day [127] were also shown to be effective. The fact that the plaque-inhibiting effect of xylitol was maintained over 2 years in the Turku feeding studies, contradicts the possibility of microbial adaptation in the oral cavity [90, 91]. The results of Turku plaque studies [92, 94] were confirmed by Plüss [612]. It should be finally noted that the plaque-reducing effect of xylitol is dose-dependent. It also depends on the frequence of use of xylitol.

Gehring [60] recently showed that the metabolism in certain rat and hamster oral microorganisms of xylitol was lower than that of mannitol or sorbitol. In another study [543] certain enterococcus strains, a micrococcus strain and a strain of *Str. mutans* were isolated under anaerobic conditions from human saliva, which metabolized sorbitol at a clearly detectable rate. Streptococcal strains isolated from rat and human oral cavity, which were able to degrade xylitol, were not found to be cariogenic in gnotobiotic rats [544]. Recently, it was shown that *Actinomyces viscosus* did not ferment xylitol [611]. When the streptococcal counts of plaque samples of subjects using either xylitol or sucrose as the sweetener in chewing gum were determined, the greatest number of negative reversals occurred in the xylitol group [620].

6.33 Metabolism of xylitol by other microorganisms

In spite of the findings with oral microorganisms, other bacteria, yeasts and molds have been found to utilize xylitol and certain other pentitols. Numerous non-oral microorganisms use both hexitols and pentitols as a source of carbon. In such cases operons for the synthesis of specific polyol dehydrogenases and permeases exist. Inducible enzymes involved have been purified [128]. Such enzymes include D-sorbitol (L-iditol) dehydrogenase of *Bacillus subtilis* grown on D-sorbitol, D-mannitol 1-phosphate dehydrogenase of *E. coli, Aerobacter aerogenes* and *Bacillus subtilis* and other organisms, all grown on D-mannitol, a D-mannitol dehydrogenase of *Lactobacillus brevis,* a ribitol dehydrogenase of *A. aerogenes,* grown on ribitol, a diol dehydrogenase of *E. coli,* etc.

For the induction to occur it is thus required that the above and possible oral microorganisms attacking xylitol possess information with regard to xylitol in the form of a Xyl or an unspecific polyol region. If such a region for xylitol does not exist, no induction should occur. The organisms may then display inducibility with regard to other polyols, but in such cases the enzyme specificity requirements may form a hindrance for the enzymic conversion of xylitol. A number of polyol dehydrogenases and polyol 1-phosphate dehydrogenases are, however, to a certain extent unspecific (for a list see [61]).

The situation would turn out more interesting if DNA transformation would force the cells, previously uninformed about a polyol, to accept new genetic traits in the form of pure DNA fragments. For example, the trait to synthesize an inducible mannitol dehydrogenase was transferred from a donor strain of *Diplococcus pneumoniae* to an acceptor strain by a process termed transformation [129, 130], a physical incorporation of DNA into the recipient bacteria. Another type of transformation also occurs with the above *Pneumococcus.* Certain mutants of these cells are incapable of forming a required capsular polysaccharide. The synthesis of this requires the formation of UDP-D-glucose, the dehydrogenation of the latter to UDP-D-glucuronate, and the formation of the polysaccharide by a transferase acting on two monomeric units. In some mutants the synthesis of the capsular polysaccharide is blocked at the level of the UDP-D-glucose dehydrogenase. When such cells are transformed with DNA from capsulated type I cells, two forms of transformed cells are found. These two transformants are represented by the expected type I polysaccharide, whereas the other type consists of binary capsulated cells. The same cell synthesizes both types I and III [131, 132]. Thus transformation would furnish oral bacteria with a means to use xylitol as a main or an alternative source of carbon.

Both of the problems mentioned previously (adaptation to physiological environments, and changes in selective permeability) are interrelated, although they can also be examined separately. Both require certain explainable changes in the genome, which lead to the synthesis of new type of specific proteins (enzymes) and capsular materials. The synthesis may presuppose physical incorporation of new genetic material, of which DNA transformation was only one example; other types are known in microorganisms.

It is understandable that in laboratory conditions certain type of 'forced' adaptation may take place more easily than in a complex and adequate growth medium (oral cavity) in which normal and required nutrients are readily available. Consequently, induction-like phenomena in cariogenic microorganisms, leading to the use of xylitol, may not necessarily take place. Even on a strict xylitol diet of 2 years, as was the case in the Turku sugar study [133], the subjects consumed other carbohydrates which provided a sufficient carbon source for the bulk of the indigenous and cariogenic oral flora. Such carbohydrates included starch, glycogen, lactose and other sugars in the form of potato, rice, bakery products, meat, milk, etc. Small amounts of sucrose were also consumed in berries, other fruits, vegetables and canned foods. The indigenous oral flora and eventual cariogenic microorganisms in xylitol-consuming subjects received the bulk of their energy from this type of carbohydrates, xylitol being nothing like the only carbohydrate present. Thus there was necessarily no reason to assume that an adaptation to use xylitol would have occurred. Such a situation was found to continue for at least 2 years in subjects consuming xylitol. It may be expected that in primates no adaptation to use xylitol is necessary as far as some glucose is also consumed. Even after 4.5-year continuous use of xylitol practically no acids were formed from it in plaque suspensions [614].

In bacteria not belonging to the prevailing species in human mouth, mutants capable of utilizing xylitol as a novel carbon have been detected. For example, a wild type of *Aerobacter aerogenes* is unable to utilize xylitol, but a succession of mutants can be isolated capable of growing on 0.2% xylitol [134–136]. A single mutation (leading to mutant X1) is sufficient to permit their growth on xylitol. The genetic change here involves the derepression of a ribitol dehydrogenase which can metabolize xylitol but can not be induced by it [137, 138]. Xylitol is oxidized to D-xylulose [139], an intermediate of the dissimilation of D-arabitol. In this case, however, the utilization of the new carbon source requires the involvement of enzymes of a preexisting pathway. The results showed that genes which belong to different metabolic systems can be directed to serve a new biochemical pathway. It is important to emphasize in this connection that the wild type of cells were not able to metabolize xylitol and that the growth medium did not contain any other carbohydrates than xylitol. The other organic ingredients of the broth were succinate and casein hydrolysate. Furthermore, the ribitol dehydrogenase involved is not very specific. Thus these in vitro findings should not be directly translated into the complex circumstances of human oral cavity.

The authors [134] of the above experiment claimed that the basis for enhanced utilization of xylitol in the second-stage mutants (X2) was an alteration in the enzyme protein, ribitol dehydrogenase. The third-stage (X3) mutants of *Aerobacter aerogenes* were fairly active in the uptake of ^{14}C-xylitol, supporting the idea that a transport system active on xylitol had become constitutive. The mutants X2 and X3 exerted lower K_m values for ribitol dehydrogenase, acting on both xylitol and ribitol, than mutant X1. The usual function of the transport system which accepted xylitol after mutation, was apparently for the transport of D-arabitol [134] (transport system for D-arabitol is present in the membrane).

Aerobacter aerogenes is able to metabolize certain pentitols as shown in figure 2. D-Arabitol is most likely concentrated by the cell without involvement of phosphoenolpyruvate system. D-Xylulose gives rise to other intermediates in the D-arabitol pathway. D-Xylulose may also be reversibly converted to D-arabitol [134]. Wu, Lin and Tanaka demonstrated an active transport system for D-arabitol, which also acts on xylitol [134]. This is shown in figure 2. The above mentioned mutant X3 produces this system constitutively and removes xylitol from the medium. In mutants X1 and X2 D-arabitol transport system is inducible. Figure 3 also includes a separate transport system for ribitol. As to xylitol, it should be emphasized that here xylitol enters bacterial cells through a transport system primarily serving other functions.

A finding related to the above case has also been made with *E. coli.* In this case wild type *E. coli* K-12 cannot grow on xylitol and it has been impossible to isolate a mutant which had acquired this growth ability [590]. However, a mutant was isolated which grew on L-1,2-propanediol as the sole source of carbon and energy. This mutant constitutively synthesized a propanediol dehydrogenase which fortuitously converted xylitol to D-xylose. The latter is normally metabolized by the cells. The enzyme responsible for the transport of xylitol into the cells is D-xylose permease. Another mutant was isolated which grew on xylitol and utilized enzymes normally needed in the metabolism of D-xylose and L-1,2-propanediol [590].

Azotobacter agilis (A. vinelandii) forms two soluble NAD-linked polyol dehydrogenases [152]. One of these, D-mannitol dehydrogenase, is induced by D-mannitol and all of the pentitols except L-arabitol. It is interesting that ribitol is an excellent inducer of this enzyme although it is not metabolized. The enzyme does not attack it. The enzyme requires that the substrate bears D-manno configuration. Consequently, D-mannitol, D-arabitol, D-rhamnitol

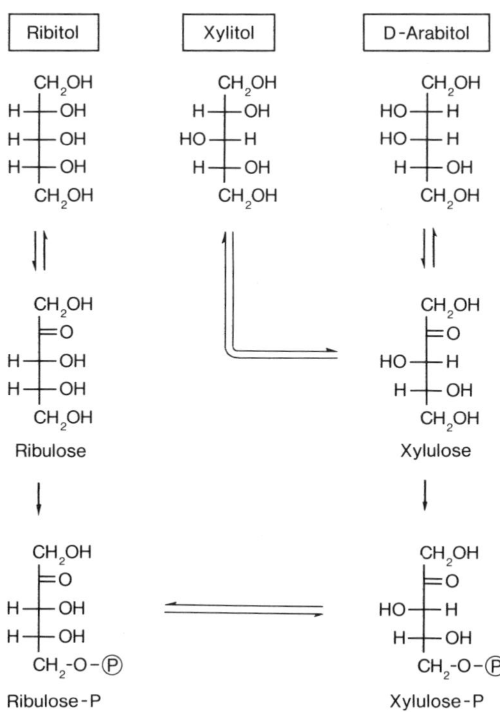

Figure 2
Enzymatic steps in the utilization of D-arabitol, xylitol and ribitol by *Aerobacter aerogenes* (according to Wu, Lin and Tanaka [134], as established by other studies [140–151]. The interconversion of D-arabitol and xylulose is mediated by D-arabitol dehydrogenase, and that of ribitol and ribulose, as well as that of xylitol and xylulose, by ribitol dehydrogenase. Note that xylitol dehydrogenase is not involved. See text for other details.

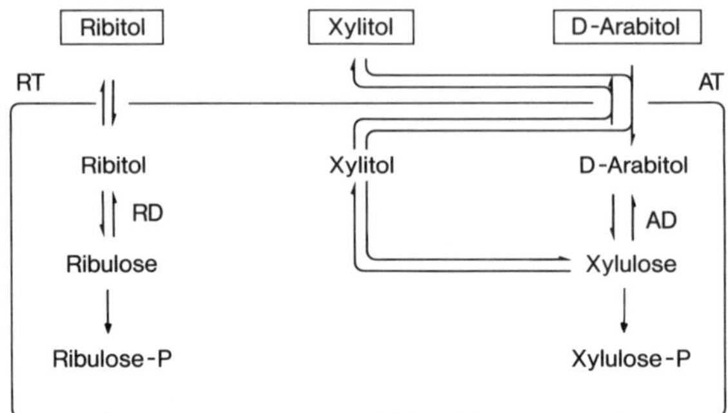

Figure 3
The pathways for transport and metabolism of D-arabitol, xylitol and ribitol in *Aerobacter aerogenes.* The system is inducible. AT stands for the transport system of D-arabitol and RT stands for that of ribitol. According to Wu, Lin and Tanaka [134].

and perseitol are oxidized. The corresponding 2-ketoses, D-fructose, D-xylulose and presumably D-rhamnulose and perseulose are reduced [152]. The second enzyme, L-iditol dehydrogenase, is only induced by polyols bearing D-xylo configuration, i.e. sorbitol and xylitol. The oxidizable substrates include L-iditol, sorbitol, xylitol and ribitol. The corresponding 2-ketosugars (L-sorbose, D-fructose, D-xylulose and D-ribulose) are reduced.

As indicated previously in the chapter on 'Aspects related to the production of xylitol' and 'Occurrence and significance of xylitol', numerous microorganisms are able to convert D-xylose to xylitol. This has been found in studies carried out, for example, by the group of Onishi and Suzuki [35–40, 47, 49] on yeasts and in those carried out by Chiang and Knight [48, 153] on *Penicillium chrysogenum.* The initial steps of D-xylose metabolism by a cell-free extract of *P. chrysogenum* involve the reduction of xylose to xylitol by a NADPH-linked D-xylose reductase, followed by NAD-linked oxidation of xylitol to xylulose.

When considering the ability of oral flora to metabolize xylitol, it is interesting to compare the substrate specificity of polyol dehydrogenases of other microbes. The comparison is shown in figure 4. A related comparison was also previously given [61]. A number of these enzymes are unspecific, thus allowing pathways for metabolism of xylitol and other polyols in these bacteria.

6.34 Utilization of xylose by microorganisms and animals

The utilization of D-xylose by microorganisms and higher organisms deserves, when considering the nature of the oral and intestinal ecological determinants, the following survey. The aldopentose xylose occurs much more wide-spread in nature than xylitol. Many woody materials, corn-cobs, cotton-seed hulls, pecan shells (*Carya pecan*, hickory), almond shells, straw and numerous other plant materials are rich in xylose units. In the putrefaction of xylan, the formed xylose is utilized by some soil microorganisms which thus are furnished by more or less constitutive enzyme systems for the transport and metabolism of xylose. This situation has incontestably continued for

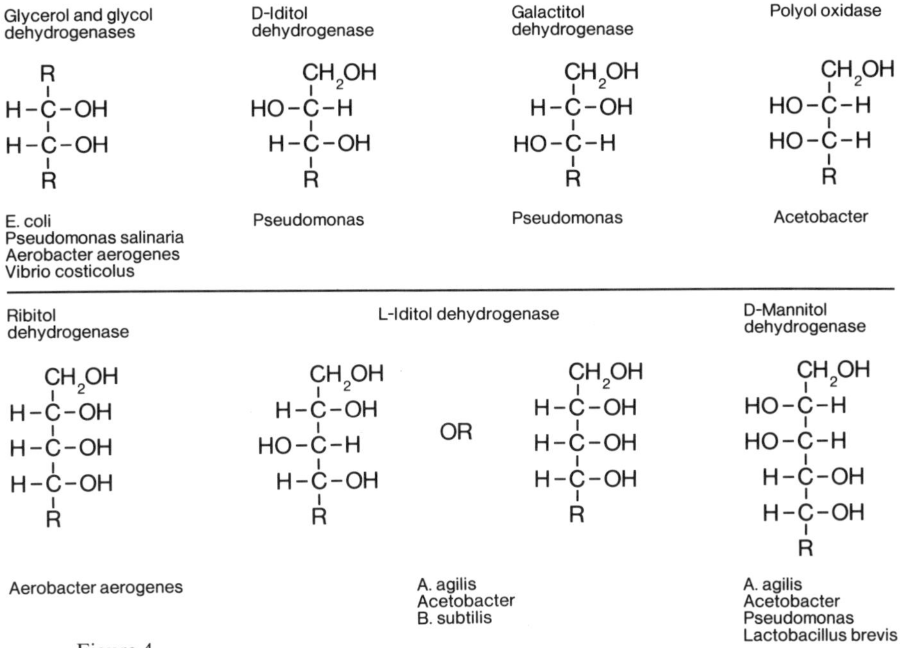

Glycerol and glycol dehydrogenases

R
H-C-OH
H-C-OH
R

E. coli
Pseudomonas salinaria
Aerobacter aerogenes
Vibrio costicolus

D-Iditol dehydrogenase

CH_2OH
HO-C-H
H-C-OH
R

Pseudomonas

Galactitol dehydrogenase

CH_2OH
H-C-OH
HO-C-H
R

Pseudomonas

Polyol oxidase

CH_2OH
HO-C-H
HO-C-H
R

Acetobacter

Ribitol dehydrogenase

CH_2OH
H-C-OH
H-C-OH
H-C-OH
R

Aerobacter aerogenes

L-Iditol dehydrogenase

CH_2OH
H-C-OH
HO-C-H
H-C-OH
R

OR

CH_2OH
H-C-OH
H-C-OH
H-C-OH
R

A. agilis
Acetobacter
B. subtilis

D-Mannitol dehydrogenase

CH_2OH
HO-C-H
HO-C-H
H-C-OH
H-C-OH
R

A. agilis
Acetobacter
Pseudomonas
Lactobacillus brevis

Figure 4
Substrate specificity of various bacterial polyol dehydrogenases. In principle according to Marcus and Marr [152].

enormously long periods during the course of evolution. Among germs being capable of using xylose or xylitol (or other polyols) there are yeasts [35–40, 44, 49, 60, 154], molds [48, 153] and bacteria [134, 152], to mention representative examples. *Monilia* and *Torula* yeasts grow well on hydrolyzed straw. Many bacteria, lactobacilli, for example, ferment xylose with the formation of lactate and acetate [155]. The aerobic yeast, *Candida utilis,* makes extensive use of a polyol pathway for the metabolism of pentoses. These cells possess a pattern of polyol metabolism similar to that of higher organisms. The cells of *C. utilis* and related species reduce xylose by NADH to xylitol. This is the first step in pentose metabolism in *C. utilis*. Xylitol is subsequently oxidized by NAD to D-xylulose [395]. The polyol dehydrogenase activity of cells grown on D-xylose is much higher than in cells grown on D-glucose. The reduction of xylose is catalyzed by an inducible enzyme. D-Xylose and xylitol are natural substrates of *C. utilis* [395].

The variation in the behaviour of microorganisms as regards their pentose and pentitol metabolism is further emphasized when species normally growing in environments low in xylitol or xylose, are considered. So, for example, the growth of luminous bacteria from dead fish was considerably inhibited by mannitol, galactitol, sorbitol, xylitol, arabinose and xylose. These sugars could not act in place of glycerol [404]. However, rhamnose (a methyl-pentose) promoted luminescence.

Sheep are also able to make use of ingested xylose, and corn-cobs provide a good cattle feed [154]. Hogs eliminate 30% of ingested xylose in the urine [154]. The pentosan digestibility has been studied with Wistar rats [156]. The arabinan fraction of hard wheat *(Tritium durum)* and millet *(Pennisetum typhoideum)* seemed to be preferentially hydrolyzed and assimilable; the xylan probably composed the bulk of fecal excretion. Ruminant animals have enzymes capable of digesting xylans. An active hemicellulase (xylanase) fraction has been obtained from cell-free supernatants of rumen contents [53].

Xylose is metabolized to a small extent by the rat [401]. About 1% of injected ^{14}C-xylose was probably converted into glycogen in the mouse liver [402] and approximately 10% was oxidized to CO_2. In the guinea-pig, 75% of injected xylose was excreted in the urine [270]. In man, 13.5% of ^{14}C-xylose was expired as CO_2; all ^{14}C-xylose excreted as unchanged sugar. No xylonic acid or xylitol was detectable in these experiments [403]. Van Heyningen [335] showed that blood xylose levels of rats fed on a 35% xylose diet remained low. The concentration of blood glucose was not much affected by xylose consumption. There may be considerable geographic variations, however, as to the metabolism of xylose by man. Low urinary excretion of xylose has been described after a 5-g oral dose in 36% of healthy Burmese subjects. A 5-g oral dose may lead to as low a urinary excretion of xylose as 5–20% of the dose. 1-hour plasma levels were then in the normal range of 13–20 mg/100 ml [537]. Consequently, under tropical conditions, low urinary excretion of xylose after an oral dose does not necessarily imply impairment of intestinal absorption. Xylose absorption by the small intestine probably includes an active process [592]. It has been pointed out that factors that influence the results of a xylose-absorption test include intestinal bacterial overgrowth, reduced xylose metabolism in liver disease, sequestration into ascites, age, and the state of renal function [592]. The 5-g dose advocated by many authors for xylose testing was considered to have disadvantages when compared to a 25-g dose [592]. Domestic and experimental animals and man seem to possess inherent mechanisms for the catabolism of certain amounts of xylans or xylose. These carbohydrates are utilized or treated by cattle in massive amounts without emersion of xylose-dependent dental caries or xylose cataract.

6.35 The possibility of xylitol adaptation in plaque

Xylose, xylans and xylitol have always existed in the environment of many microorganisms which have thus made use of these carbohydrates. In the human oral cavity both the indigenous and cariogenic flora have actually never been seriously faced by any unavoidableness to deliberately elaborate specific enzymes for the breakdown of xylose or xylitol, because the oral ecosystem has simply comprised largely water-soluble hexoses or other soluble sugars based on six-carbon structures. These have formed the most easily and abundantly available water-soluble source of energy and carbon in the oral cavity. These carbohydrates have been the natural ecological determinants. In addition to soluble hexoses and related sugars, also amino acids have been easily available nutritional determinants. Only occasionally more or less viable mutants or strains may exist in dental plaque – such have been demonstrated – which are equipped with enzyme systems attacking xylitol or other pentose-based sugars. As a result of the presence of xylitol in most berries, other fruits and plant material, plaque has always been exposed to certain low concentrations of dietary xylitol. In spite of this, practically no xylitol fermentation takes place in dental plaque, although possibilities for this have existed. It is obvious that human oral microorganisms will not become adapted to use xylitol to any significant extent as long as hexoses (glucose) are offered at present concentrations in the human diet. The previously presented considerations of molecular genetics support this idea.

6.4 Effect of xylitol on enzymes

6.41 Inhibition

The ability of xylitol to inhibit certain enzyme reactions may be manifested as an inhibition of the translocation of sugars across bacterial cell membranes. Various mechanisms involved in the transport exist, for example:

- Phosphotransferase system.
- Permease system.
- Specific sugar-binding proteins.
- Free diffusion.

The transport of xylitol across the bacterial cell membrane does not most likely take place through free diffusion. Regardless of the nature of the mechanism, it at certain stages obviously presupposes *binding* to membrane components. Binding is involved in the function of practically any simple or conjugated protein. Regardless of whether xylitol is translocated or not, its binding may inhibit some of the above transport reactions of sugars required by the cells. In case xylitol were transported to a certain extent, it might inhibit intracellular enzymic conversations of other monosaccharides. In many cases, however, the selective permeability of membranes usually forms a protective barrier against compounds not involved in normal metabolism, thereby hampering translocation of xylitol. These possibilities have not been studied intensively, and so far not at all with cariogenic microorganisms. A few interesting examples of enzyme inhibition can be mentioned here.

D-Xylose isomerase. Inhibition of D-xylose isomerase of *Lactobacillus brevis* by pentitols and D-lyxose has been reported [160]. Of the pentitols xylitol and D- and L-arabitol inhibited competitively, but there was a considerable difference between the inhibition constants determined. K_i was 0.0027 M for xylitol and 0.13 M and 0.146 M for D- and L-arabitol, respectively. Thus xylitol was clearly a more potent inhibitor than the other polyols tested.

In the above case of the xylitol-inhibited D-xylose isomerase it was suggested that the substrate and the inhibitor combine with Mn^{2+} ions through the C_3–C_5 hydroxyl groups of the molecule by a similar mechanism. This can be considered obvious, because the C_3–C_5 regions in D-xylose, xylitol, D-arabitol and D-lyxose resemble each other. Two stereochemical requirements were thus revealed: One is for binding (met by the D-lyxose structure); the other is for activity. Among the above-mentioned compounds xylitol most closely resembles the structure of D-xylose. It is possible that the differences in the structure at C-1 must thus have a significant effect on binding to the isomerase [160]. The above suggestions were based on open-chain structures [160]. The majority of pentoses in physiological solutions occur as furanoses in which there is no longer any reactive carbonyl group in the molecule [162]. If the enzymes discussed act on open-chain molecules, the explanation above is adequate. In case the enzyme acts on ring forms, the inhibitory effect of xylitol should be partly explained on other grounds. It may be important to note that certain pentose isomerases also catalyze the isomerization of D-glucose [161]. Enzymatic isomerization of pentoses has been shown in a variety of bacteria, including D-arabinose isomerases of *E. coli*, D-xylose isomerase of *Lactobacillus pentosus, Pseudomonas hydrophila, Pasteurella pestis,* and *L. brevis*, and L-arabinose isomerase of *L. pentosus, Aerobacter aerogenes* and *L. gayonii* (for a list of references, see [160]).

D-Glucose isomerase. Xylitol also potently inhibits the D-glucose isomerase of various *Actinoplanes* species. The relative activities measured for this enzyme in the presence of various polyols were: 35, 70, 80 and 100 for xylitol, D-sorbitol, D-galactitol and D-mannitol, respectively (at 0.1 M). D-Ribitol did not inhibit the reaction [163]. The results showed that xylitol inhibited more strongly than the other polyols. This Mg- and Co-activated enzyme also acts on D-xylose and D-ribose.

The size or length of a carbohydrate molecule frequently determines the productivity of its collisions with the active site of enzyme proteins. There may be numerous reasons why a shorter molecule may not act as a reactive substrate or inhibitor. One reason may be related to the ability of the reactive molecule to displace the hydration layer at the active site of an enzyme [164]. An ice-like structure of water surrounds the enzyme proteins in an aqueous solution. It is obvious that the interaction of any molecule with enzymes must to some extent be determined by the water envelope. Consequently, the ability of a sugar molecule to bind to an enzyme may depend on whether the molecule can disrupt the water structure. Interaction would thus involve displacement of water. It can be postulated that in certain cases the xylitol molecule, due to its shorter length, fails to displace water molecules to a sufficient extent at the active site of an enzyme acting on six-carbon sugars. Depending on the case binding may still occur, resulting in enzyme inhibition. The existence of the water layer may in some cases also explain why xylitol does not inhibit enzymes in which the active site would otherwise enable the molecule to land in a right position. The configuration at various carbon atoms also determines the productivity of collisions. For example, while D-fructose is readily metabolized by plaque bacteria, L-sorbose is not [611].

6.42 Specificity requirements of enzymes

Evaluation of the metabolism of xylitol in the dental plaque requires the consideration of enzyme specificity. Enzymes of cariogenic and other oral microorganisms may be able to discriminate against xylitol and other polyols which fail to bring into play any of the catalytic factors enumerated by Blow and Steitz [165] as follows:

- Covalent intermediate (xylitol would not form covalent ES complexes with certain enzymes).
- Entropic effects.
- Strain effects (no sufficient strain in the xylitol molecule would form when it binds to enzyme).

In the determination of enzyme specificity, according to Blow and Steitz [165], the following four situations may thus be important: (a) *Lock and key.* Exact fitting of glucose, xylitol or some other sugar into the binding site discriminates against incorrectly shaped sugars or those with other unfit properties. Ribonuclease is an example of this hypothesis. (b) *Productive binding.* An appropriate substrate binds only in the required orientation to promote catalysis. Chymotrypsin serves as an example. (c) *Strain.* An appropriate sugar binds in a way to create sufficiently strong binding interactions. This fit provides energy to strain the sugar, or bonds between sugar units, towards the conformation of a transition state. Lysozyme is an example. (d) *Induced fit.* A sugar of correct structure is able to trigger a special change in the enzyme. The change orients the mutual position of the catalytic groups of the active site and causes it to become enzymatically active. This mechanism is said to account for high specificity [165]. A number of enzyme inhibitions by xylitol may be explained on the basis of (a), (c) and (d) effects. The induced fit principle offers a few interesting examples in the metabolism of carbohydrates, which are briefly dealt with below [166, 167]. These considerations may help to understand the behaviour of xylitol among other dietary carbohydrates.

According to Koshland some small molecules are not attached to the enzyme surface, while their larger analogues are bound. Thus β-glucosidase usually acts on glucosides but not on 2-deoxyglucosides. The catalytic groups of the enzyme have little or no effect on such compounds, despite the fact that similar adsorbed substances are reactive. These and related findings led to the formulation of the hypothesis of induced structural correspondence [166, 167]. This postulates that (a) the substrate causes material changes in the enzyme geometry, since it penetrates into the active center, and that (b) precise orientation of the catalytic groups is necessary for catalytic action. It further postulates that (c) the substrate induces the proper orientation by the changes it produces in the enzyme geometry. The reaction occurs only when the effective groups of the enzyme have definite three-dimensional arrangement.

This was proved by Koshland with β-amylase which acts on the terminal groups of amylose but not on other glucoside bonds of the polysaccharide (and not on cycloamylose). It can be expected that in the case of several enzymes acting on hexoses and their derivatives, a smaller molecule (xylitol or xylose) may not be attached onto the enzyme surface as strongly as the actual substrates. Xylitol and xylose would thus be poor substrates for such enzymes. If the enzyme is inhibited by xylitol, binding occurs, but the polyol may not meet all requirements of the enzyme (concerning, for example the length of the molecule).

Literature has also described other cases which may elucidate the possible mechanism of the effects of xylitol on oral bacteria commensal to man. Glycosidases, other than those described by Koshland, also form an interesting example [168, 170].

Glycosidases form a group on enzymes which are concerned in the main with substrates with no charged groups. As in xylitol, all the determining groups in glycosides are most often either hydrogen atoms or hydroxyl groups. Consequently, the specificity of glycosidases must be partly determined by the pattern in which these groups are arranged. The glycosidases seem frequently to be specific for a particular monosaccharide ring. The attached glycone group may in some cases have a more or less marked influence, because some glycosidases are specific for the aglycone (nucleosidases).

When the substrate contains no charged groups, as is the case with most simple dietary sugars, forces other than electrostatic must be responsible for the attraction between an enzyme and substrate. In the case of enzymes acting on sugars in which the main feature of the combining structure is an arrangement of hydroxyl groups, combination can be explained in terms of hydrogen bonding. Thus hydrogen bonding may also form the basis of the competitive enzyme inhibition and specificity requirements touched upon in this chapter. Many competitive inhibitors are substances with acidic or basic groups, but non-ionizing molecules such as sugars often act as competitive inhibitors for enzymes acting on similar molecules.

In general, the interchange of hydrogen and hydroxyl on any single carbon atom of a glycoside substrate is sufficient to prevent the action of the corresponding enzyme [168–170]. For example, β-glucosidases do not usually act on β-D-mannosides or α-D-glucosides. In the case of epimers involving the C_1, C_2 and C_4 of the aldohexose ring, separate enzymes usually exist for each structure. Substitution on the hydroxyl groups of the sugars usually has a profound effect; with β-glucosidase any substitution on C_2, C_3 or C_4 completely prevents hydrolysis. Replacement of the $-CH_2OH$ group attached to C_5 by $-H$, i.e. conversion to a β-D-xyloside, produces 200-fold reduction in rate. More or less specific β-xylosidases will act more readily on the pentose substrates.

Brain hexokinase serves as another interesting point of comparison. It is competitively inhibited by D-xylose [171]. This substance is identical with the respective substrate except for one smaller change, i.e. the substitution of a $-H$

for a $-CH_2OH$ on C_5. The inhibition is caused by the absence of a $-CH_2OH$ group, which results in a decrease of the number of contact points (in the form of hydrogen bonds and hydrophobic interactions) between the enzyme and substrate. The shorter molecule, D-xylose, does not lead to reactive complexes. Consequently, enzymes which normally are active toward hexose-based sugars, are often inhibited by pentose-based ones.

6.43 Specifically on sugar transport through bacterial cell membrane

The use of sugars as an intracellular source of energy requires translocation of the sugar molecule through the cell membrane. Transport of xylitol in dental plaque is most likely very rare. In cases when it is transported through bacterial cell membranes several different mechanisms may be involved [157–159]. Sugars may be taken up by specific binding proteins (there are galactose-binding proteins, for example) or by facilitated diffusion (like frequently with glycine). Features of constitutivity and inducibility are involved in transport. The role of phosphoenolpyruvate in the P-transferase system has been elucidated [172–175]. The system catalyzes the transfer of phosphate from P-enolpyruvate to various carbohydrates according to the following overall-reaction:

sugar + P-enolpyruvate → sugar-P + pyruvate.

The phosphate donor for the reaction is P-enolpyruvate and various sugars act as acceptors. All sugars studied by the group of Kundig [172–175] were of D-configuration and glycosides were pyranosides. With the exception of fructose, all sugars studied [172–175] were phosphorylated in position 6 (glucose, mannose and the corresponding hexosamines). Certain P-transferase systems catalyzed, however, the phosphorylation of fructose in position 1. Certain parts of the transferase system are constitutive, certain are inducible. In some cases the system includes two enzymes, Enzyme I and Enzyme II, and a heatstable protein (HPr). HPr and Enzyme I are constitutive and most Enzymes II are inducible [172–175]. It is interesting that preparations from some strains of *E. coli* grown under *non-induced* conditions still phosphorylated galactose and mannose at low rates. The possibility that this low constitutive activity towards galactose in isolated membrane preparations may be involved in induction of the galactose operon, has been suggested [157].

These suggestions and results can be used to claim that inducibility of enzymes involved in carbohydrate translocation is more easily understood and more readily demonstrable if the phenomenon is based on even a very low constitutivity (or on the existence of a particular sugar operon). If, however, the degree of constitutivity of an enzyme system (including its operons) required in the translocation and utilization of xylitol is nil, no use of xylitol is possible, without the appropriation of new genetic material, or the involvement of a mutation.

6.44 Dextran-antidextran system

The reactions between antibodies and antigens are not very far from the complementary nature of many interactions between enzyme and substrate. It has been suggested that immunization with glucosyltransferase might prove a measure in preventing dental caries [176]. On the other hand, disposal of bacterial antigens from the oral cavity readily takes place with immunoglobulins [177]. These and numerous other studies suggest that partially immunological factors are involved in dental caries. It has been suggested that the antibody structural genes control the immune response to dextran [178]. About 10 years earlier it had become evident that oligosaccharides of the isomaltose series are competitive inhibitors of dextran-antidextran precipitation [179]. Oligosaccharides of this type (with isomaltose pattern) may be formed in plaque, for example, by dextranases. From the dental point of view the dextran-antidextran precipitation is an advantageous process. It can be postulated that inhibitory oligosaccharides are formed to a high extent on a sucrose diet. Persons consuming a xylitol diet may produce such compounds only to a lower degree. This is possible because xylitol decreases the quantity of plaque which may contain inhibitors of the dextran-antidextran reaction.

6.5 **Cariogenicity of polyols**

The bacterial metabolism of certain polyols was touched upon in sections 6.32 and 6.33. The following findings should also be considered. Xylitol has not been shown to induce any significant acid production in plaque. No microbial adaptation to utilize xylitol has been found in dental plaque. On the other hand, sorbitol is attacked by adaptive enzymes and this polyol is slowly fermented by oral microorganisms [180–183]. In the isolated dental plaque sorbitol did not cause a pH-decrease in contrast to sucrose [184–186], but still produced some lactic acid although less than sucrose [598]. Radiotelemetric pH-recording of interdental plaque areas revealed a moderate acid formation from sorbitol with a pH value of 6 or higher [116, 599]. Sorbitol may thus be considered 'slightly acidogenic'. In contrast, xylitol causes no drop of the pH at all and may thus be approved as 'non-acidogenic'. An in vitro experiment on plaque suspensions showed that xylitol prevents the sucrose-induced acid production more effectively than sorbitol, when the sugars were tested at fairly low 0.6% concentrations [539]. In vitro experiments suggest that sorbitol can be regarded as strongly acidogenic [614], depending on the subject studied and the length of the incubation time.

Animal experiments have claimed that sorbitol is less cariogenic than sucrose [187–190], but human clinical trials have been unable to demonstrate any significant difference between the cariogenicity of sucrose and sorbitol [191], although such could be expected to exist [225].

A recent series of experiments carried out by Ainamo et al. [192] in Finland on dental students showed that the chewing of a chewing gum sweetened with sorbitol did not increase or decrease the normal growth of dental plaque. The cleansing effect of sorbitol-containing chewing gum was also tested on the plaque formed during 4 days with no oral hygiene. The chewing of ten pieces of sorbitol gum during a time period of 2½ hours did not significantly reduce the plaque scores. Frostell [193] and Gülzov [194] have come to the conclusion that acid production from sorbitol may increase in time in the oral cavity. Other independent experiments showed that the use of xylitol-containing chewing gums inhibited plaque growth by 30–55%, although the subjects refrained from oral hygiene [90–94, 126, 127]. Xylitol chewing gum was effective even when compared with a group of non-chewers. Based on the above findings and the information of chapter 6, the non-cariogenic nature of xylitol may be regarded as an understandable fact.

The caries-reducing effect in rat fissure caries was demonstrated with both xylitol and sorbitol [195]. Other animal experiments have shown the excellent anticariogenic properties of xylitol [196, 197, 600, 601]. The controversal reports on xylitol-induced rat caries [198] have been partly explained as resulting from methodological reasons. The authors [198] later showed that 10 or 20% xylitol in the diet did not increase caries incidence in rats when compared to the control diet (corn starch) [548]. The caries-reducing effect of xylitol in rats was recently verified [622].

The strong caries-inhibitory effect of xylitol was further elucidated in the Turku sugar studies [99–101]. Based on the clinical, radiographic, biochemical and microbiological data, the authors of the Turku studies suggested that xylitol has a curative, remineralizing effect on dental caries in man. The effect was extended to enamel and dentine caries of many subjects studied. Effective doses were found to be even less than 6–7 g per day in the form of chewing gum. The effect also comprised fissure, smooth surface and approximal caries. A virtually complete eradication of dental caries was thus possible using a natural carbohydrate as a sweetener in chewing gum or in the diet. Details of these studies have been published [99–101].

Although many workers indicated that sorbitol is slightly cariogenic in rats and hamsters [187, 189, 190, 195, 199, 200], there were still several partly contradictory considerations about the benefits of sorbitol in dentistry. The promising findings about the effects of the application of sorbitol solutions to human plaque [184, 193, 201], or to monkey plaque in situ [202], which failed to produce the rapid and extensive falls in pH found when sucrose or glucose solutions were similarly applied, led to the obvious misunderstanding that sorbitol would be fully safe in dentistry. On the one hand, it was shown by Clark et al. [203] that the rate of acid production from sorbitol by whole saliva was not influenced by the consumption of three tablets of sorbitol daily for 4 weeks. The dosage of sorbitol was fairly low in this short-term study. Frostell [204] demonstrated a marked increase in the capacity of plaque to produce acid from sorbitol during 2- or 6-month periods of sorbitol consumption. It can be concluded that while sorbitol seems to be fairly low-cariogenic in monkeys *(Macaca irus)* [205], the same can not be expected in human subjects which clearly differ from monkeys in having a somewhat reduced salivary defence mechanism against caries [206, 244]. If the interdental plaque pH is used as criterion, short-term mouth rinse studies with 10% solutions of xylitol or sorbitol in man reveal small, but noticeable differences between these polyols since sorbitol in contrast to xylitol decreases the pH by roughly 1 pH-unit [116, 405, 598, 599]. According to the current knowledge, acid production should not, however, be regarded as the only criterion. The available literature indicates that sorbitol is a mildly cariogenic carbohydrate. Caries is formed in its presence.

Information about the cariogenicity of other polyols is scant, apart from mannitol which is slightly cariogenic or non-cariogenic. Glycerol seems to be non-cariogenic in man. A recent study, based on an intraoral cariogenicity test [549], claimed that maltitol and Lycasin® (a complex mixture of hydrogenated hydrolysis products of starch) will be non-cariogenic if the mechanism of lesion initiation in the test and natural caries are similar [550]. It is obvious that most natural hexitols behave almost like sorbitol and that most natural pentitols share the dentally beneficial qualities of xylitol. However, the physiological, organoleptic and other properties of xylitol seem to place it first among present carbohydrate sweeteners in human diet.

The history of polyol research in dentistry has been short; scant in the beginning and impetuous in the recent years. There have also been certain misconceptions related to the molecular properties of sorbitol and xylitol. Certain scientists have maintained that all polyols should have a similar effect on dental caries. This is naturally an oversimplification, as can be concluded on basis of chapters 2 and 6.3. Polyols clearly exhibit selective effects in the oral cavity.

6.6 Facilitation of mineralization by xylitol

The consumption of xylitol in the Turku sugar studies decreased the incidence of dental caries in man by more or less than 100% when compared to the consumption of sucrose. The exact degree of reduction depended on the type of test person and the way the results were expressed. The results were obtained by clinical and radiographic examinations. The reduction of caries incidence was accompanied by remineralization of caries lesions [99–101]. Earlier dental and physicochemical literature has occasionally touched upon related phenomena, but without the consideration that the consumption of a dietary natural carbohydrate would lead to this type of effect. However, from the theoretical point of view even sucrose can lead to remineralization. This is possible during the short period immediately after sucrose intake before the plaque starts ferment sucrose. Many sugars may increase the salivary flow rate and pH. Higher pH values favour remineralization better than low values. The sugars differ from each other, however, with regard to fermentability and osmotic properties.

Saliva is an electrolyte solution which is saturated with regard to hydroxyapatite, the main inorganic constituent of teeth and bones. Consequently, normal saliva is also saturated with regard to enamel, because dissolution of this structure commences only when pH in the environment has been lowered [207–210]. The data obtained when studying the rate of the salivary flow and the concentration of calcium and phosphate under various stimuli give rise to interesting suggestions about the possible remineralization and curative, therapeutic and prophylactic effects of xylitol in dental caries in man [207, 211–215]. It is a well-known fact that most sugars and certain carboxylic acids (as citric acid) increase the rate of the flow of saliva. For example, when studying the rate of the parotid flow and calcium concentration, human subjects displayed higher values with increasing concentrations of a citrate rinsing solution [215]. The top value was caused by an acidic candy. This property is to a certain extent also shown by cariogenic sugars (like sucrose). However, with xylitol practically no decrease in the salivary or plaque pH results, whereas, following ingestion of sucrose, plaque pH values as low as 4.0 can be observed. The saturation of saliva with regard to calcium phosphate (or hydroxyapatite) maintains a situation in which the equilibrium between the tooth mineral and solution (saliva) favours mineralization. There are both theoretical and experimental grounds for believing that the solubility product of phosphate and calcium ions at different pH in the mouth remains more favourable when xylitol is consumed than sucrose. It would be necessary in this context to consider the possible relationship between the intake of xylitol and secretion of human anionic peptides in saliva, which inhibit the precipitation of calcium phosphate.

In a study of Grøn [216] an explanation was sought for the behaviour of salivary phosphate in an experimental system (how calcium and phosphate are rendered non-ultrafiltrable in saliva; phosphate retention). Calcium phosphate precipitation has been suggested as the reason for phosphate retention. Grøn proposed that strongly basic amino acids, with their positively charged centers, would provide binding sites for salivary phosphate. The suggestion was obviously based on a previous observation on the presence in parotid saliva of a small, strongly basic molecule with 60% of the amino acid residues being lysine, histidine, and arginine [217]. Leach et al. [218] found arginine, histidine and lysine to account for about 12% of the amino acid residues in duct saliva. Phosphate interactions with positively charged α-amino groups have been postulated to form the association between hydroxylapatite and collagen [219]. A pH-rise factor of human saliva was shown to be a C-terminal arginine peptide [547]. Guanidino groups could thus have several important physiological functions in saliva. Kleinberg et al. [613] have later provided more information about the arginine peptide, sialin.

Xylitol consumption was found to increase the concentration of basic amino acids (arginine, lysine, histidine) and glycine in whole saliva [220]. Although the functional integrity of these amino acid residues in collagen and the above-mentioned basic compound differs from that of free amino acids in saliva, an electrostatic attraction between phosphate (e.g. HPO_4^{2-}) and the mentioned amino acids would anyway be possible. Phosphate, bound to various organic molecules, is settled down with calcium in mineral-deficient enamel and dentine. Due to the nature of the deposits, simultaneous pigmentation of varying degree may occur. This is often found with arrested (rehardened) caries lesions and such was also detected in xylitol-consuming subjects in the 2-year feeding study in Turku. These findings are related to those made by Bibby [221]. According to Bibby, the continuous acquisition of various organic material (lipids, polysaccharides, amino acids, etc.) of saliva, food or bacterial origin, modifies the form of early enamel caries, contributing to the increased caries resistance and pigmentation of old teeth [221].

In conjunction with studies in which Grøn [209, 222] showed that saliva is often saturated in regard to $CaHPO_4 \cdot 2H_2O$ (dicalcium phosphate)[2]), it was also concluded that saliva contains a non-ultrafiltrable factor which may be adsorbed onto enamel powder, synthetic hydroxyapatite and salivary cellular residues [209]. This factor may be present in plaque in which it may inhibit the hydrolysis of $CaHPO_4$-hydroxyapatite compound. In this way the factor counteracts the action of fluorine. It would be erroneous to treat the mechanism of remineralization on purely inorganic grounds, omitting biological reactions. Therefore, the involvement of organic groups in the above mentioned factor should be considered. Related to the above considerations, it is interesting that xylitol-consumption keeps fluorine in the tooth structure better than sucrose and fructose [223].

Rehardening of caries lesions is a generally known biological phenomenon. The chemical conditions and factors contributing to it may not be fully understood at the present time. Recent experiments in connection with the Turku sugar studies indicate that the chewing of a xylitol-containing chewing gum induces a higher salivary pH and at least a comparable or higher P_i, Ca^{2+} and HCO_3^- ion concentration than chewing a corresponding sucrose, fructose or sorbitol product, or the gum base [568]. Consequently, carbohydrate sweeteners clearly exhibit selective effects on the chemical properties of saliva. Independent studies carried out after the Turku trial [224] also suggest that mere mouth rinses with a 10% xylitol solution lead to a temporary increase of salivary pH values which are more alkaline than those obtained with sucrose, many other sweeteners, or water. It is thus possible that bicarbonate ions are also produced at a higher rate into saliva in the presence of xylitol than many other sweeteners. A recent in vivo remineralization study of Koulourides [225] comes to the conclusion that of all sweeteners tested, only xylitol and xylose can be considered non-cariogenic. The higher sweetness and other advantageous properties of xylitol compared to xylose gives the former the preference over xylose in dietetics.

2) Different salivary preparations may be saturated in different ways with regard to hydroxyapatite, tricalcium phosphate, octacalcium phosphate, dicalcium phosphate, etc., depending on gland, subjects (man or experimental animals), stimulation, etc.

The claims about remineralizing and curative effects of xylitol in the Turku sugar studies have been occasionally criticized because a non-sweetener group was not included in the 2-year feeding study [99, 100] and in the 1-year chewing gum test [101]. For the 2-year study this suggestion is unrealistic. In the 1-year chewing gum trial the original idea was to compare two sweeteners, sucrose and xylitol, in chewing gum. Furthermore, a close examination of the clinical and radiographic data of both trials clearly shows that the finding could not be explained in other ways. The subjects of the 2-year feeding trial consumed approximately 50–70 g xylitol daily for 2 years. A few persons used more than 200 g daily, whereas in many cases lower than 50 g amounts were taken in. A number of xylitol group subjects did not adhere to the dietary regimen very strictly, but occasionally consumed certain amounts of sucrose. In spite of these less cooperative persons, a reduction of approximately 90% in caries incidence was achieved for the whole test group [99, 100]. This must indicate a percentage reduction higher than 90 or 100% with persons closely following the instructions given. Such a correlation was obvious. The cariostatic and curative qualities of orally consumed xylitol were further revealed in the 1-year chewing gum trial [101]. In spite of continuous and daily use of sucrose, small amounts of xylitol (even less than 6–7 g daily) were able to practically eradicate caries in 6–12 months.

Mastication certainly plays an important role in the functional regulation of the salivary glands not only in rats [226] but naturally in human subjects as well. The beneficial effects [101] of the use of xylitol-containing chewing gum can not be explained only on the basis of increased mastication, because subjects using sucrose chewing gum masticated to a similar extent [101]. On the other hand, most studies have shown that xylitol does not significantly increase the salivary flow rate when compared to sucrose. However, even small increase in the flow rate may lead to advantageous results.

The possibility that xylitol consumption stimulates the secretion of HCO_3^- ions from the salivary glands, has not been tested adequately, but preliminary studies have shown that the use of a xylitol chewing gum may induce a higher secretion of HCO_3^- ions than many other sweeteners [568]. This type of proposed increase in the concentration of salivary HCO_3^- ions most likely would not have any connection with increase in cerebrospinal fluid HCO_3^- ion concentration, reported in dogs given mannitol [406]. Dogs form a very poor research model in dentistry, as far as human dental caries and use of polyols are concerned.

It should be finally noted that the formation of calcium-containing microcrystals in dental plaque is a normal reaction. During the course of sucrose consumption the development of these crystals is locally retarded or hampered by acid production. When xylitol does not decrease plaque pH, the formation of the microcrystals renders remineralization of carious lesions possible.

6.7 **Effects of xylitol on plaque and saliva in Turku sugar studies**

The subsequent paragraphs list a number of biochemical phenomena observed in the oral cavity of human subjects placed on long-term xylitol diet.

6.71 Amino acid composition of oral fluid

Microorganisms deprived of a required nutrient, e.g. sucrose or glucose, often exhibit elevated intra- and extracellular proteinase activity. The increase of the activity may in some cases be explained as an enzyme induction, glucose repression or as an appearance of latent proteinases. The extracellular enzymes are needed to produce substrates for energy-requiring processes. This results in increased amino acid concentration in the extracellular medium. Both phenomena, viz. the elevation of the proteinase activity and the concentration of most amino acids in human whole saliva occur on xylitol diet [91, 220]. The last mentioned phenomenon is also encountered on fructose diet [220]. After the detection of the increased proteinase activity in vitro in cultures of *Streptococcus mutans* [109, 111], the respective enzymes have been partially purified [227]. Because the consumption of a xylitol diet strongly reduces the incidence of *Str. mutans,* the proteinases found in elevated amounts in whole saliva may have derived from other bacteria of indigenous flora. The xylitol-dependent increase of basic amino acids in whole saliva may play a role in the remineralization of carious lesions. The consumption of a xylitol diet did not increase the concentration of urea in whole saliva [220]. On the contrary, the urea concentrations showed a tendency to decrease. Urea is perhaps the most rapidly degraded nitrogenous compound in saliva producing ammonia, CO_2 and an elevated pH in plaque. These considerations are in accordance with those of Kleinberg [613].

6.72 Invertase-like activity

The earlier observations [88, 89] about the lower whole saliva and plaque invertase-like activity of persons on xylitol diet compared to consumption of sucrose diet were verified in the 2-year feeding study [90–94]. The term invertase-like here stands for enzyme(s) liberating reducing sugars from sucrose; the enzyme activity was determined as reducing sugars. Consequently, in the enzyme assay on whole saliva some glycosyltransferase activity was also included. Furthermore, no evaluation of enzymes attacking the liberated reducing sugars in the crude reaction mixtures could be carried out. However, the net result was to a high extent caused by a decrease of true invertase(s), as shown by chromatography and biochemical experiments [91]. The effect partly resulted from lowering of invertase-producing bacterial flora in the presence of xylitol, and partly from the absence of proper substrate (sucrose) needed in invertase induction. A lower intake of xylitol, as in the form of chewing gum, does not necessarily affect invertase or glycosyltransferase activity in human whole saliva [230].

Invertases of *Str. mutans* [84, 86] and plaque [88, 89] have been claimed to be inducive. On the other hand McCabe et al. [85] stated that the invertase is induced by sucrose in *Str. sanguis*, but is constituent in *Str. mutans*. It was earlier suggested that the consumption of a xylitol diet may lead to the induction of plaque invertases [88–90]. As background for the above findings one should consider the nature of true induction. Enzyme induction is usually regarded as a phenotypic response of a culture to the presence of the appropriate chemical signal in the environment, resulting in an increase of a 1000-fold or more in the synthesis of enzymes so controlled. Many catabolic enzymes are induced by their substrates. A series of enzymes of a biochemical pathway can also be sequentially induced by the intermediates of the pathway. The rate of synthesis of many catabolic enzymes is altered considerably depending on the presence of other carbon and nitrogen compounds. Catabolic repression is followed when high growth rate and the availability of rapidly metabolized sugar or other carbon compound lead to a decrease in the rate of synthesis of inducible enzymes. It is possible that in the case of plaque invertases [88–90] no true induction is necessarily involved, but rather a control of enzyme activity and level by inorganic phosphate or other compounds [87, 228].

Str. mutans invertases have been suggested to be intracellular [86]. However, extracellular invertases also occur in plaque [229]. Lysis of cells liberates additional invertases into the extracellular phase of plaque.

6.73 Plaque lactate

Lactic acid is said to represent the bulk of the carboxylic acid pool in plaque, although the importance of acetate and propionate has also been emphasized [122, 231]. Certain other carboxylic acids may also affect enamel dissolution [232]. Pyruvate also occurs in plaque in considerable amounts, most likely representing the bulk of the keto acid pool of plaque extracellular phase [91]. Xylitol consumption clearly lowered the amount of lactate in plaque, whole saliva and their component fractions [91]. This reduction largely resulted from the simultaneous inhibition of both total plaque growth and incidence of acid-forming bacterial species. However, when lactate was expressed versus plaque wet weight, the ratio was lowest in the xylitol group, indicating that the concentration of lactate in plaque was also reduced [91]. The lower quantities of plaque lactate following ingestion of xylitol lead to insignificant dissolution of enamel compared to the use of sugars which are readily fermented. This situation prolongs the remineralization periods occurring in mouth compared to the consumption of sucrose-rich diet.

6.74 Oral glycosidases

Plaque samples of subjects consuming a xylitol diet exhibited somewhat higher (bacterial) glycosidase activity toward p-nitrophenyl-β-D-galactoside than subjects consuming sucrose or fructose [91, 233]. The significance of this phenomenon is partly unknown. It may be related to microorganisms' search for nutrients when the cells are deprived of sucrose. The increased glycosidase activity may be a reflection of secretion of specific salivary glycoproteins required in the formation of the acquired pellicle, a thin organic layer covering and protecting the teeth. The bulk of the overall glycosidase activity in the human oral fluid with certainty originates from bacteria. In this sense it may be interesting to view the claim, frequently presented, that plaque only restrictedly contributes to the enzyme content of gingival crevice fluid [234]. A xylitol diet, however, clearly lowered gingival exudate glycosidase activity. The bulk of the exudate glycosidase activity may be derived from serum and tissues but involvement of bacterial origin cannot be totally excluded. These enzymes may occur in exudate depending on the severity of gingival inflammation. The xylitol-dependent increase in the fucosidase activity in whole saliva may be associated with the finding that the salivary glycoproteins occur in plaque matrix devoid of their fucose (or other sugar) residues. The protein moieties so formed are capable of forming Ca-precipitates in the presence of phosphate. Consequently, the increased glycosidase activity may be indirectly associated with remineralization.

6.75 Peroxidase activity

The consumption of a xylitol diet increased the salivary lactoperoxidase activity as determined in human whole mouth saliva [223]. This enzyme has been shown to act as a natural and physiological defence mechanism in the mouth [235, 236]. Several interesting possibilities can be considered as causes of the xylitol-dependent increase of lactoperoxidase:

(1) Macromolecular activation.
(2) Activation by smaller molecules.
(3) Direct or indirect effect of oral flora on the enzyme activity.
(4) Increased de novo synthesis or conversion of inactive proenzyme to active enzyme in the salivary glands.
(5) Stimulation of the emptying of acinar cells.

There are several examples in the literature of cases in which non-enzymatic macromolecules affect catalytic processes [237]. Succinoxidase is potently activated by serum albumin, gelatin, casein, peptone and globin. Peroxidase is activated by keratin and gelatin. The activation may be due to the presence in these macromolecules of hydrophilic groups in aqueous media, although the proteins themselves are often insoluble. The macromolecules may form weak enzyme-mimicking complexes with phenolic compounds, like guaiacol, used as the lactoperoxidase substrate, and with Fe(III) ions. The effect of keratin and other proteins was, however, partly excluded in the Turku studies because the same increase in enzyme activity was achieved after chromatography on CM- or DEAE-cellulose as well [223]. The increased lactoperoxidase activity was found in the chromatographic fractions which most likely did not contain keratin. Xylitol per se does not activate lactoperoxidase at the concentration used [223].

Another possible reason for the elevated salivary lactoperoxidase activity, i.e. the effect of varying concentrations of activating SCN^- or iodine ions, was also considered unlikely [223]. The increased activity was also observable after molecular permeation and ion exchange chromatography which would have abolished any such ion effects, as the small SCN^- ions would most likely appear in the last fractions at least in molecular permeation chromatography. It was also concluded that catalase, acting on H_2O_2, cannot be considered a source for the differences observed. There was, however, an apparent correlation between lactoperoxidase activity and the concentration of SCN^- ions in subjects who used high amounts of xylitol ([223] and table 11 in [91]). Furthermore, SCN^- ions may bind to proteins [591], suggesting that verifying studies should be carried out.

Xylitol-promoted microbial changes may also indirectly lead to higher lactoperoxidase activity. In such a speculative case the bacterial flora prevailing on sucrose diet would affect the lactoperoxidase-SCN^-(or I^-)-H_2O_2 system maintaining an enzyme activity level normally observed in human whole saliva. Because xylitol diet strongly influences the oral flora, the enzyme system would accordingly have a higher (more uninhibited) activity. Bacteria anyway contribute to the normal lactoperoxidase activity of whole saliva by furnishing the enzyme system with hydrogen peroxide, a substrate of the enzyme.

The fourth possibility, an increased de novo synthesis, or formation of active enzyme from an inactive proenzyme in the glands, can be considered more likely. The chemical message for the production or liberation of elevated amounts of salivary lactoperoxidase can in principle pass as in the case of amylase. Increased consumption of protein-rich diet was suggested to lead to an increase of the concentration of gastrin which would reduce the adenylate cyclase activity in the salivary glands. This in turn prevents amylase from appearing in higher amounts in saliva [242]. A rich carbohydrate diet would have an opposite effect. The fifth possibility should also be considered.

The selective effect of various food components on the chemical composition of human saliva has not been sufficiently emphasized. One of the first papers indicating the above-mentioned selectivity appeared in 1925 by Walker (quoted by Afonsky [238 and 239]). The activity of salivary α-amylase was increased by a predominantly carbohydrate diet. This type of enzyme changes are quite natural and physiological. Various chemicals also cause different effects in saliva. For example, the composition of the carbohydrate moiety of the glycoproteins of the submaxillary saliva of dogs was shown to change upon the dose of pilocarpine [240, 241].

The inorganic composition of saliva is also changed pending the chemical properties and concentration of the compounds taken in. Calcium, bicarbonate, hydrogen and phosphate ions, for example, behave in this way. Xylitol displayed a selective effect by causing a decrease in the concentration of salivary F^- ions, which could not be encountered on sucrose diet. Other studies have recently emphasized the fact that different dietary regimens significantly modify protein and amino acid levels in young rat parotid glands. These modifications may serve as a mechanism by which diet may influence oral health [551]. All these examples indicate the importance of the effects of diet on the function of the salivary glands.

It is also possible that sugars undergo a rapid diffusion-like phenomenon from the oral cavity to salivary glands through mechanisms not so far well known. Such a mechanism could also explain the lactoperoxidase effect.

Stimulation of the cervical sympathetic ganglia on intravascular injections of epinephrine cause a marked increase in the permeability of the submaxillary salivary gland of cats and dogs to certain non-electrolytes [243]. When using a series of carbohydrates with known molecular size, it was possible to estimate the size of the 'intercellular gaps' in the gland necessary to allow passage of sugar molecules from blood into saliva. Bradykinin or kallidin strongly increased the saliva/plasma ratios for sucrose and glucose. The polypeptide effect on salivary gland permeability was considered to be due at least in part to the release of epinephrine from certain storage sites in the body. Bradykinin, however, was still shown to alter the permeability of the submaxillary salivary gland to sucrose and glucose when effects of epinephrine on permeability were blocked. For the molecular radius of glucose a value of 0.34 nm was given. Inulin with an average radius of 1.47 nm did not penetrate the gland [243]. According to these authors the maximal opening of the intercellular gap was approximately 2.5 nm, allowing for the passage of molecules (including xylitol) with a radius of 1.2 nm or less. It can be argued, however, that mere physical passage without actively participating membrane components is more unlikely.

In order to elucidate the mechanism of the xylitol-dependent increase of lactoperoxidase activity, a series of experiments was commenced on monkeys in the National Institues of Health [206]. Short-term (3 days) consumption of xylitol increased the lactoperoxidase activity of both submandibular and parotid saliva, cannulated directly from the glands, as compared to sucrose consumption. Thus the monkey experiment verified the results of the Turku studies. In the next series of experiments monkeys were subjected to gastric intubation with xylitol solutions (2.5 g per day and animal) for 3 days [244]. Preliminary results indicate that the dosage was too low to cause any clear effects on salivary or lacrimal lactoperoxidase activity. As a result of experimental arrangements it was not possible to conclude if there was any difference between xylitol and sorbitol.

For an effective action of the salivary lactoperoxidase it is not necessary for the enzyme to be bound onto enamel, although it is natural that such interaction takes place and that the bound enzyme is in a stable and enzymatically active form [545, 546]. The human salivary lactoperoxidase is also rapidly adsorbed by the cells of *Str. mutans* [569] and most likely by other oral microorganisms as well.

A recent study [606] on human parotid saliva showed that a single dose of xylitol-containing candy increased the lactoperoxidase activity of human parotid saliva [606], whereas in another study [607] such an effect was not observed with whole saliva. These two reports were not in disagreement, as the experimental conditions were very different. In the former [606] study parotid saliva was collected during a period of 15 minutes after a 15-minute stimulation of saliva with seven to fifteen fruit pastils (gum pastils). In the other study [607] whole saliva was collected by paraffin stimulation immediately (0–15 min) after an initial 2-minute stimulation with a single xylitol-containing chewing gum. In the latter study the stimulation was short and the amount of xylitol used was very small. The effect of xylitol on the salivary lactoperoxidase thus depends on the way of administration, amount and form of the xylitol product used, and the time allotted to protein synthesis in the salivary glands.

As a result of the low absorption rate of xylitol, oral administration of this polyol does not reach such concentration levels in blood, which would affect the activity of myeloperoxidase and other leukocyte enzymes [603].

The recent finding that thyroxine may act as a regulator [617] of the iodinase and peroxidase enzyme(s) of submaxillary gland, should be considered when explaining the xylitol-associated lactoperoxidase effect.

6.76 Other effects of xylitol

A number of less pronounced chemical changes have been found to take place in the mouth as a result of xylitol consumption. The list below includes a few examples.

- Xylitol consumption increased the ratio of proteins to carbohydrates in dental plaque [91].
- Xylitol consumption slightly increased plaque aspartate transaminase activity [91]. This effect was most likely related to the general increase of protein metabolism in plaque and whole saliva during xylitol consumption. Other similar phenomena were the increase of amino acid concentration [220] and proteinase activity in whole saliva [91].
- Whole saliva amylase activity was shown to have decreased as a result of xylitol consumption [91]. The enzyme assay method was, however, not very specific. Subjects consuming sucrose harboured more starch-splitting bacterial enzymes in plaque, which also attacked hydrolyzed starch used in the amylase assay. Hence no idea about the effect of the true salivary amylase was obtained. Recent experiments on monkeys indicate that short-term consumption of xylitol diet increases true salivary amylase activity, tested with cannulated parotid saliva [245].

There was thus an apparent discrepancy between the human [91] and monkey [245] studies, which could be explained by the use of whole saliva in the human experiments.

- Xylitol consumption decreased the salivary hydroxyproline concentration [91, 220] when compared to sucrose- or fructose-consumption. Saliva and plaque contain hydroxyproline-degrading enzymes, but it is equally likely that the results indicated lessened hydrolysis of collagenous structures in the mouth during the use of xylitol.
- Xylitol consumption decreased the concentration of ionizable fluorine in whole saliva (determined at the normal pH of whole saliva; pH 6.8–7.4) [223]. A greater part of fluorine remained in the teeth in the xylitol group due to a low rate of dissolution of the hydroxyapatite structure.
- Gingival exudate was shown to contain less enzyme activity of certain aminopeptidases [246–248], peroxidase [223] and glycosidases [233] in the xylitol group when compared to the sucrose and fructose groups. From the periodontal point of view, this may be considered an advantageous situation. These enzyme changes resulted partly from inhibition of plaque by xylitol [90–94, 126, 127].
- Other effects of the consumption of a xylitol diet have been described in detail elsewhere [90–94, 133, 220, 230, 233, 570].

Regarding all these chemical and biochemical changes or reactions it has to be emphasized that the saliva and plaque samples were not necessarily collected after immediate oral loading of the subjects with xylitol. Samples obtained immediately after a high xylitol intake may behave in a different way.

6.77 Effect of long-term use of xylitol-sweetened chewing gum

Since the publication of the clinical and biochemical results of the 1-year chewing gum trial [101, 230], certain additional information on pooled whole saliva representing the xylitol and sucrose groups, each of approximately fifty subjects, has been obtained. Although a part of the findings mentioned were obtained on pooled samples, they may be interesting in light of later similar trials. Table 4 compares properties of the whole saliva pools and individual samples obtained from the chewing gum study. The data are not complete because most of the values shown were obtained with samples collected at the end of the study only. These are shown merely to encourage further studies in this field, because it is possible that salivary ions and mucins are selectively affected by xylitol.

Table 4 indicates that the whole saliva samples differed to a certain extent depending on the type of sweetener in the chewing gum. The saliva samples were obtained in the mornings of the collection days without any immediately preceding stimulation with chewing gum. This certainly led to the fact that no differences were found between the two groups in the activity of lactoperoxidase and invertase-like enzymes. The daily dosages of xylitol were most likely too small (3–15 g, on the average 6–7 g) to cause detectable changes of enzyme activity in samples collected in the morning. As the use of xylitol-containing chewing gum was shown to reduce the incidence of dental caries very effectively, it is still possible that the explanation to this could be partly found in the chemical properties of saliva. Therefore, the

Table 4

Chemical properties of centrifuged whole saliva of subjects who chewed approximately 4.5 pieces of chewing gum daily over 1 year. Each piece contained 1.5 g of either sucrose or xylitol. The sweeteners in the chewing gums were the only main variables in the study. All values were determined on samples collected after 1 year use of the chewing gums. Details were described in [101] and [230].

Property	Xylitol group	Sucrose group	Remarks
Proteins (mg/ml)	0.97 ± 0.20	0.98 ± 0.22	Mean and S.D. of 50 subjects[1])
Lactoperoxidase (U/l)	18.1 ± 7.4	18.1 ± 8.2	Mean and S.D. of 50 subjects[1])
Enzymatic liberation of reducing sugars from sucrose μmol/(min × mg)	12.8 ± 9.6	15.3 ± 11.8	Mean and S.D. of 50 subjects[1])
Behaviour of suspensions after suspending freeze-dried solids in water[2])	Larger flakes and particles	Fairly homogenous suspension	Pooled samples (50 subjects)
Proteins (mg/ml)	0.80	0.83	Pooled samples (50 subjects)
Lactoperoxidase (U/l)	14.0	13.0	Pooled samples (50 subjects)
Inorganic phosphorus (mg/ml)	0.131	0.125	Pooled samples (50 subjects)
Calcium (mg/ml)	0.17	0.12	Pooled samples (50 subjects)
SCN^- ions (mg/l)	104.5	100.0	Pooled samples (50 subjects)
Sulphate (μg/ml)[3])	12.0	10.0	Pooled samples (50 subjects)
Ionized iodine ($M \times 10^6$)	4.1	1.6	Pooled samples (50 subjects)
Sialic acid (μmol/ml)[4])	0.04	0.05	Pooled samples (50 subjects)

1) The means and standard deviations are from [230]. Analyses were made immediately after collection.
2) Individual samples were pooled after the first three chemical assays, freeze-dried, stored 3 months at −20 °C, suspended in 10 ml water, and centrifuged for analyses. Analyses on pooled samples were thus made 3 months after saliva collection and the results are given for the original volume of saliva, not for the freeze-dried concentrate.
3) After hydrolysis of saliva in 0.5 N HCl and at 100 °C for 2 hours.
4) After hydrolysis of saliva in 0.2 N H_2SO_4 and at 80 °C for 1 hour.

chemical values shown in table 4 for the saliva pools may to a certain extent reflect such possible advantageous properties of saliva collected from subjects who used xylitol chewing gum.

6.8 Conclusions: outlines of the xylitol effects

The informations provided in previous chapters lead to the following suggestions about the chemical and physiological background of the effect of xylitol in dental caries. The three properties of xylitol shown below should be considered simultaneously.

(1) Unsuitability of xylitol as a bacterial substrate. Because xylitol is a pentitol, human oral microorganisms, and particularly the cariogenic species, have not been able to use xylitol as a source of energy and for production of acids and harmful extracellular polysaccharides in plaque. Therefore, xylitol has not been a cariogenic carbohydrate in man. Human cariogenic microorganisms more effectively use carbohydrates based on six-carbon structures. Such compounds have been decisive ecological determinants in the human oral cavity. After xylitol intake human oral fluid (whole saliva) and plaque remain more alkaline than after sucrose intake.

(2) Xylitol may form complexes with Ca^{2+} and certain other cations. The complexation may interfere with the adhesion of bacterial cells in plaque.

(3) Relationship between colligative properties of xylitol and exocrine gland function. Both peroral administration and gastric intubation of xylitol increase, in a dose-dependent manner, the secretion of certain specific proteins and inorganic ions from salivary glands. Depending on the dose of xylitol, these proteins include lactoperoxidase and amylase as well as non-enzymatic substances. The latter most likely comprise glycoproteins and mucins which normally protect the mucous membranes of the oral cavity and give rise to the acquired pellicle, a thin organic layer increasing the resistance of tooth surfaces. Preliminary information of Turku sugar studies indicates that various acidic groups of salivary mucins appear in different ratios depending on the chemical structure of the sweetener.

In addition to the salivary glands, other glands of the alimentary canal also respond to the presence of xylitol. The function of the secretory cells involved includes production of mucopolysaccharides and glycoproteins which protect the mucous membranes in the intestines and the stomach. All parts of the alimentary canal which are influenced by the friction caused by passing materials in the canal, may be affected by xylitol administration. Xylitol may cause changes in mucin, mineral and water secretion and diffusion in the mucosa. These reactions are partly caused by the colligative or osmotic qualities of xylitol and the changes take place within the normal physiological range.

The mastication of xylitol products may lead to an osmotic effect in the oral cavity as well. Here the mucous membranes may react much like those in other parts of the alimentary canal. Consequently, changes in water and electrolyte balance may partly explain the xylitol effect in dental caries. The changes may include an increase in the concentrations of calcium, phosphate, HCO_3^-, and OH^- ions of the oral fluid. These responses are fairly quick, taking place in 1–2 minutes after starting to chew a chewing gum. Xylitol may also reach the minor salivary glands in low amounts directly from the oral cavity. The minerals involved are thus partly derived from saliva, but a certain portion most likely also diffuses through the mucous membranes in the mouth. There is thus a slight loss of tissue calcium and certain other ions. Depending on the xylitol dose, this calcium is partly or almost totally reabsorbed in the lower parts of the alimentary canal. Other sugars also share these properties, but with xylitol the smaller molecular size may lead to more favourable dental effects. The increased pH and the higher concentrations of calcium and phosphate ions render more effective transient remineralization periods possible. Because the intestines become adapted to tolerate very high concentrations of xylitol, it can be questioned, if this also happens with regard to oral mucous membranes. Such adaptation has not so far been described.

At the molecular level the exact mechanism, which would explain in detail the chemical reactions of the mucous membranes in the alimentary canal after xylitol intake, is not known. The following considerations should be taken into account:

- Human organs, apart from liver, do not possess any significant capacity to metabolize xylitol. Many tissues do not metabolize it at all.
- As a result of the size, configuration and symmetry of the xylitol molecule it is fairly rapidly resorbed at the apex of the intestinal villi in which xylitol is then distributed in the epithelial layer. Here, or in the connective tissue of the villi, there is, however, a delay in the further translocation of xylitol. This stage may correspond to the generally found slower rate of absorption of xylitol compared to glucose.

The last-mentioned phenomenon is not unique to xylitol, but all osmotically effective and slowly absorbed compounds behave in a related way. If a higher quantity of xylitol is administered enterally, there is a transient accumulation of xylitol in the intestines. The increased concentration induces an osmotic effect and water is drawn to the epithelial layer. A larger portion of this water is translocated into the intestinal lumen, resulting in an osmotic diarrhea of short duration. Similar osmotic effects may take place in the oral cavity, but the cellular mechanism naturally differs from that suggested for the gut wall.

One is entitled to compare the effects of xylitol in the oral cavity with those displayed by hypertonic solutions of osmotherapeutics used in clinical practice. Hypertonic solutions of sorbitol, mannitol, xylitol, with addition of glucose, fructose, dextran, and other compounds, show a diuretic effect. The losses of water are naturally accompanied by losses of Na^+, K^+ and Cl^- ions.

The volume of the extracellular space and plasma is therewith affected. A predominantly normotonic excretion of electrolytes is finally established. A single administration cannot thus produce a lasting dehydration. In spite of the differences between urinary and salivary excretion and secretion, the osmotic properties of xylitol remain the same. It is suggested that the xylitol effects in the oral cavity can be more easily understood if the effects in the other parts of the alimentary canal are used as a point of comparison. Studies elucidating these relationships are being carried out. The following finding also suggests that such a relationship should be considered: It is generally known that calcium absorption is increased by many sugars such as lactose, xylose, sorbitol and xylitol. The involvement of a Ca-sugar complex has been proposed, but later results do not support this hypothesis [605]. Rat experiments showed that xylose was not absorbed [605]. This resulted in an extensive secretion of water out of the cell into the lumen. As xylitol is absorbed fairly slowly, a related, but minor secretion of water occurs after the intake of this sugar alcohol.

Consequently, the effect of xylitol in dental caries seems to be dual. The effect is partly local and related to bacterial physiology, and partly systemic. The systemic effect may also be partly dual, including two separate mechanisms. One may operate through the route stomach-circulation-salivary glands. The other involves a more rapid transmittance of a physico-chemical message to oral mucous secretory cells and possibly to the major salivary glands. The effects of xylitol on exocrine glands are actively investigated in several laboratories.

7. Xylitol in dietetics and medicine

7.1 Discovery of enzymes of xylitol metabolism in mammals

The detection of L-xylulose in the urine of non-pentosuric humans [250–252], guinea-pigs [251] and rats [252] indicated that this ketopentose may be a normal metabolite in mammals. This idea was supported by the discovery of a very active enzyme system for the reduction of L-xylulose to xylitol [253]. It was also found that D-xylulose 5-phosphate, rather than D-ribulose 5-phosphate, is the ketopentose substrate of transketolase in the 6-phosphogluconate pathway [254, 255]. Simultaneously, new mitochondrial enzymes linking metabolically the two enantiomorphic forms of xylulose were studied in guinea-pig liver [249].

In these and subsequent studies, carried out largely by Touster and Hollmann, it was further established that guinea-pig liver mitochondria catalyze the reduction of L-xylulose to xylitol. Reversibility of the reaction xylitol↔ketopentose was confirmed [253]. The enzyme responsible for this reaction was located in the insoluble part of mitochondria. The reaction required pyridine nucleotides. The specificity of the enzyme was very high, because no activity toward several ketoses and polyols tested, except for xylulose and xylitol, did occur [253]. This high specificity thus differentiated the enzyme from other enzymes known at that time to catalyze the interconversion of ketoses and polyols. The 'L-xylulose-xylitol enzyme' of guinea-pig mitochondria was found to need NADP and its high specificity was reconfirmed [256]. In addition to the involvement of this type of highly specific enzymes, Hollmann and Touster [256] emphasized the multiplicity of NAD-dependent liver polyol dehydrogenases. These dehydrogenases catalyze the interconversion of several polyols and ketoses. It is likely that the use of guinea-pig as one of the first test objects was a fortunate choice and to a certain extent speeded up the understanding of xylitol metabolism in man, because both species require α-ascorbate in the diet. On the other hand, xylitol and ascorbate are closely related to each other through the glucuronate-gulonate pathway, or glucuronate-xylulose cycle, established subsequently [257, 258]. In liver perfusion tests the utilization of L-xylulose was slower with hamster, rat and monkey than with guinea-pig.

Hollmann [259] purified satisfactorily two enzymes in guinea-pig liver mitochondria: a NADP-xylitol(L-xylulose) dehydrogenase and a NAD-xylitol(D-xylulose) dehydrogenase. The NADP-dependent enzyme catalyzed the reactions xylitol ⇌ L-xylulose. The NAD-dependent enzyme was a Zn-containing SH-enzyme, capable of oxidizing the following polyols: L-iditol, D-sorbitol, xylitol, D-glycero-D-glucoheptitol, ribitol, allitol and L-threitol. Thus the required configuration was that of a D-erythro-1,2,4-polyol. Of all substrates xylitol had the lowest value of K_m (5.96×10^{-4} M) [259], indicating a high affinity for the enzyme among the polyols tested.

A recent study [597] suggested that the L- and D-xylulose reductase catalyzing the reversible NAD-linked reduction of D-xylulose and the NADP-linked oxidation of xylitol proceed with opposite chirality of hydrogen transfer to the coenzyme. This is an important suggestion as this type of enzymes have so far been found to catalyze hydrogen transfer from and to the coenzyme invariably with the same chirality.

The above and certain other studies, referred to later, have resulted in the availability of information about xylitol metabolism, which has up to now been summarized and discussed at several symposia of which three main are as follows:

(1) 'Metabolism, Physiology, and Clinical Use of Pentoses and Pentitols', Hakone, Japan, August, 1967 (published by Springer-Verlag, New York, in 1969) [290].
(2) 'Sugars in Nutrition', Nashville, Tennessee, U.S.A., October, 1972 (published by Academic Press, New York, in 1974).
(3) 'Monosaccharides and Polyalcohols in Nutrition, Therapy and Dietetics', Basle, Switzerland, June, 1975 (published by Verlag Hans Huber, Bern, in 1976).

A recent symposium in Würzburg (Symposium II über Zuckeraustauschstoffe; Dtsch. zahnärztl. Z. *32*, Suppl. 1) mostly dealt with sugar substitutes in the prevention of dental caries. The first symposium revealed, for example, the following findings:

- Xylitol was shown to be less lipogenic than sucrose.
- Intravenously administered xylitol was considered effective and safe in young and old human subjects, both with regular metabolism as well as with diabetes, liver disease and after surgery.
- Xylitol augmented adrenal steroidogenesis, preventing suppression under steroid treatment.
- Xylitol was considered useful in suppressing ketosis resulting from amino acid diets in pediatrics.
- Xylitol had a favourable effect in reducing bilirubin levels in infantile jaundice.
- The above findings could be largely explained by the fact that xylitol was found to be rapidly oxidized in animal and human tissues, including red blood cells in which it prevented hemolysis. Xylitol is oxidized to xylulose in entering the pentose phosphate pathway.
- Xylitol was found to restore glycolysis in diabetic rat liver slices without the use of insulin.

- Xylitol was found to be of value in the treatment of diabetes and other ketogenic metabolic situations such as the postoperative phase.

Results of the second symposium are touched upon in more detail in this review. Certain findings presented in the latest symposium in Basle have contributed to the knowledge about xylitol. However, most of the information has been partially repetitive and can be found in original papers referred to in this review. A basis for these conceptions about xylitol metabolism in mammals was pioneered by the biochemical enzyme studies mentioned in the beginning of this chapter.

7.2 Aspects of the metabolism of xylitol in mammals

7.21 General features

Sugar alcohols are biochemically very close to aldose and ketose sugars. Consequently, xylitol and other related polyols in the present survey can also be treated as sugars. This practice also consists with certain chemical encyclopedias which regard all crystalline and sweet carbohydrates as sugars. Many polyols are as sweet as sucrose or glucose and they are to a large extent metabolized as the two latter ones. Chemically, there is at least as big a difference between a disaccharide and glucose as between glucose and sorbitol, for example. A noticeable error is to classify sorbitol as a sugar and xylitol as a polyol [571].

There are profound differences between the rates at which sugar alcohols are metabolized in animal tissues. Administered sorbitol, ribitol and xylitol are usually very effectively metabolized. Of these xylitol seems to be metabolized by man more efficiently than the two other polyols. The utilization of D-(+)arabitol, L-(-)arabitol and dulcitol (galactitol) is very poor [257]. Mannitol is absorbed and metabolized in man [261, 262]. In case of mannitol the route of administration is said to be very decisive, oral doses being utilized much better than parenteral doses [261]. If this and certain other considerations would also have been kept in mind with regard to administration of xylitol, confusing reports about biochemical abnormalities with xylitol would have been more readily understood [263–265]. However, xylitol seems to be better tolerated than any other sugar alcohol, regardless of the route of administration.

Various mammalian species have been found to handle sugar alcohols either by (a) oxidation to the corresponding ketose, or (b) oxidation to the corresponding aldose. Phosphorylation of the polyol to polyol phosphate, frequently encountered in certain microorganisms, has not so far with certainty been detected in mammals. In summary, the two important sugar alcohols, sorbitol and xylitol, are interconverted in human and many other mammalian tissues to corresponding enantiomorphic forms as shown below:

L-xylulose ⇌ xylitol ⇌ D-xylulose
D-glucose ⇌ sorbitol ⇌ D-fructose

All these reactions, except for the glucose-sorbitol interconversion, are catalyzed by polyol dehydrogenases which need NAD or NADP as coenzymes. The xylitol-D-xylulose and sorbitol-D-fructose reactions are most likely catalyzed by the same enzyme or by closely related enzymes [266]. The glucose-sorbitol step usually needs a NADP-dependent aldose reductase.

While oral administration of even high amounts of xylitol seems to be safe, it is necessary to emphasize the importance of due consideration of the turnover rate in intravenous nutrition with xylitol. This naturally concerns the intravenous nutrition in general regardless of the substrate used. A third type to investigate xylitol involves perfusion tests with isolated organs, tissue slices or blood cells. The overall effects of xylitol in these systems certainly differ to a certain extent from those encountered when the whole body is involved. It is likely that a number of erroneous conclusions have been made on the basis of such investigations. Misunderstanding happend also when it has been observed that functional effects may result from the excess production of polyols such as sorbitol or galactitol in the nervous or in vascular tissues [572]. Oral as well as intravenous administration of xylitol at a moderate rate will not lead to such phenomena.

7.22 Glucuronate-xylulose cycle

In the genetic abnormality essential pentosuria, L-xylulose appears in urine in fairly high concentrations. Attempts to show whether this sugar, when incubated with tissue preparation, would give rise to other common sugars or sugar phosphates, have failed. Instead of these reactions L-xylulose was reduced to xylitol by a specific NADP-dependent polyol dehydrogenase, a largely soluble cytoplasmic enzyme [267]. This enzyme was first thought to be present only in mitochondria [256]. The intramitochondrial xylitol dehydrogenase is probably not involved in the metabolism of exogenous xylitol [269].

Formed xylitol is then subsequently dehydrogenated by a NAD-requiring L-iditol dehydrogenase (EC 1.1.1.14; also known as sorbitol dehydrogenase), yielding D-xylulose. This product is then phosphorylated to D-xylulose 5-phosphate which enters the pentose phosphate pathway [270]. These and other findings [271] led McCormick and Touster [270] to formulate the glucuronate-xylulose pathway, the main features of which are shown in figure 5. Partially with the aid of this pathway ^{14}C-labelled xylitol is metabolized to glycogen and CO_2. Thus the products of the metabolism of xylitol are channelled mainly into glycolysis which is located in the cytoplasm, as are the other enzymes required in the complete xylitol metabolism. Additional information needed in the outlines of the glucuronate-xylulose cycle was obtained from studies of Burns, Kanfer and Ashwell [260].

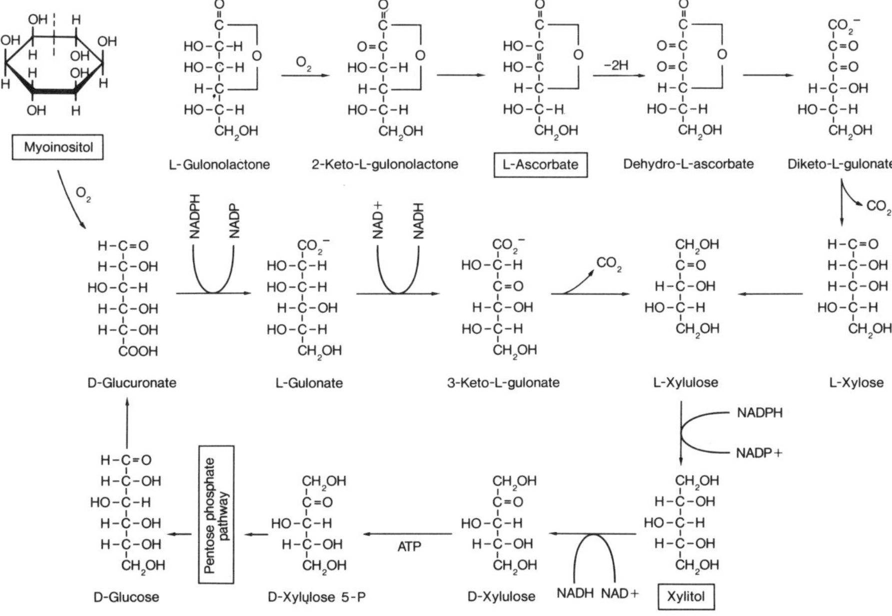

Figure 5
The glucuronate-xylulose cycle as described by Touster [266].

The glucuronate-xylulose cycle is responsible for the synthesis of L-xylulose and ascorbic acid. All primates, guinea-pig, flying mammals and insects have a genetic enzymatic defect which makes the microsomal oxidase incapable to yield ascorbate from L-gulonolactone [266, 272].

Certain restricted human populations have a defect in L-xylulose utilization. This defect is rare and it occurs recessively in Arabs and Jews only [273, 274]. According to Touster [266] the cycle has been demonstrated in all animals tested, but its main functions have not been elucidated with certainty. It has been suggested [266] that the early steps of the pathway would be linked to the synthesis of nucleotide sugars which are needed in the formation of glycoproteins, mucopolysaccharides, glycolipids and glucuronides. The last mentioned are detoxification products of the body. The cycle obviously plays an important role in the production of mucin-like compounds for the whole alimentary tract. Most animals also synthesize ascorbic acid via this pathway. The cycle could also serve a transhydrogenation role because several steps utilize NADPH while others produce NADH [266]. It has been estimated that the total carbohydrate flux via the cycle is between 5 and 15 g daily in man [275], corresponding to several percent of the daily oxidized carbohydrates. The cycle is enhanced by certain drugs and hydrocarbons [275, 276].

The two xylitol dehydrogenases occurring in animal livers, also occur in red blood cells [277]. Pentosuric subjects seem to have a normal NAD-linked dehydrogenase, but a mutant NADP-linked enzyme with a reduced capacity for the reduction of L-xylulose to xylitol [278, 279]. The pentosuric erythrocyte enzyme has a higher K_m for NADP than the normal enzyme. This suggests that the enzyme is synthesized in pentosuric subjects, but in a modified and less active molecular form.

To understand the rapidity at which liver cells are able to handle xylitol and convert it into other intermediates of the carbohydrate metabolism, the permeability of the liver cell membrane for xylitol (and sorbitol) should be emphasized. Usually more than 80% of xylitol is metabolized in the liver. Both xylitol and sorbitol are rapidly taken up both by isolated perfused rat liver and liver in vivo experiments. Other cell membranes, for example, in the lens, are clearly more impermeable (c.f. chapter 7.28 later). After having entered the hepatic cells, the oxidation to D-xylulose by the NAD-xylitol dehydrogenase takes place rapidly. Heavier loading with xylitol can thus increase the NADH-NAD ratio in liver [280, 281]. The liver D-xylulose kinase, performing the next phosphorylation step to D-xylulose 5-phosphate, has been partially purified and characterized [282].

After entering the pentose phosphate pathway, D-xylulose 5-phosphate is finally converted into fructose 6-phosphate and glyceraldehyde phosphate. According to Touster and Shaw [257] three xylitol molecules produce two molecules of the hexose phosphate and one molecule of the triose phosphate. Because fructose 6-phosphate can be easily converted into glucose and glycogen, the physiological behaviour of dietary xylitol can be understood. Resulting glyceraldehyde phosphate is naturally metabolized to glucose and glycogen as well, but it can also yield lactate through a few enzymatic steps via phosphoenolpyruvate and pyruvate [283]. That a main product of the metabolism of xylitol in liver is glucose has been confirmed by independent laboratories [284–287].

It is not precisely known whether administration of a particular intermediate of the pathway increases the function of the cycle as a whole, thus producing advantageous clinical results or adverse reactions. Certain recent findings suggest [223, 244], however, that there is a possible relationship between xylitol administration and increased synthesis of certain mucopolysaccharides and/or glycoproteins of exocrine glands, particularly salivary glands. There is a possibility that xylitol administration directly or indirectly increases the efficacy of detoxification reactions and increases the resistance of mucous membranes. Enteral xylitol most likely stimulates the formation of salivary mucins and glycoproteins (see paragraph 7.27).

7.23 Assessment of certain metabolic and toxicological studies

Jacob et al. [288] provided evidence that most of the xylitol subjected to oxidation to D-xylulose and phosphorylation to D-xylulose 5-phosphate in liver, is rapidly converted into glucose. This glucose is subsequently liberated to the circulation. Only small quantities of lactate are formed. In addition to liver and erythrocytes, at least kidney and adipose tissue [283, 289, 290] have been shown to take up xylitol and incorporate it into various metabolites. Xylitol incorporation into total lipids and xylitol oxidation to $^{14}CO_2$ were not affected by the presence of sorbitol. Thus there was no competition between sorbitol and xylitol in the experimental system involved [283]. Froesch and Jacob concluded that enzymes other than sorbitol dehydrogenase are responsible for the metabolism of sorbitol and xylitol in adipose tissue, and that xylitol metabolism in this tissue is very small and probably without any physiological significance. It was further shown that xylitol had no influence on the metabolism of fructose by adipose tissue [283].

Rat liver perfusion tests [281] also showed that xylitol metabolism was predominantly cyanide-sensitive, indicating an obligatory involvement of mitochondrial respiration for removal of reducing equivalents from the cytosol. Xylitol metabolism was not controlled by the activity of NAD-xylitol dehydrogenase but by the rate of reoxidation of NADH. A flavin-linked shuttle, identified with the α-glycerophosphate shuttle, was postulated [281]. The α-glycerophosphate shuttle operated by virtue of the high cytosolic α-glycerophosphate concentration produced during xylitol metabolism in the perfusion tests. Involvement of a NAD-dependent malate-aspartate shuttle was also suggested, because glucose formation from xylitol was sensitive to aminooxyacetate and β-hydroxybutyrate [281].

In other series of experiments it was shown that even in the presence of insulin, sorbitol and xylitol were extremely rapidly converted into glucose by the rat liver [283, 291]. However, the metabolism of xylitol and sorbitol (and fructose) seems to differ from that of glucose with respect to the initial steps by which these sugars enter the glycolytic pathway of liver. These steps are naturally independent of insulin and may have led, according to Froesch and Jacob [283], to the misunderstanding that the entire utilization of these polyols and fructose is also insulin-independent [289]. The utilization of glucose in the rat was shown to be insulin-dependent in all tissues investigated by Froesch and Jacob [283, 291], i.e., liver, muscle and adipose tissue. Because glucose homeostasis is well maintained in rat, the rapid conversation of xylitol or sorbitol does not produce significant hyperglycemia.

The conclusion of Froesch and Jacob [283] that xylitol and sorbitol do not offer any advantages for parenteral nutrition contradicts other and more frequent statements about their benefits (for example, Lang [289]). Froesch has later very strongly claimed that sugars like sorbitol and xylitol are experimental tools only without any place in parenteral nutrition [564]. He claimed, on the basis of a rat experiment, that these polyols would not be insulin-independent [564]. Froesch and Jacob [283] also warned about the danger of lactic acidosis if large amounts of fructose, sorbitol or xylitol are administered. It is possible, however, that one is here to a certain extent dealing with the usual discrepancy arising when interpreting results of animal tests and transfering the results obtained to human conditions. Most authors and the overwhelming majority of publications seem to agree that xylitol, sorbitol and fructose are brought into the metabolism independently of insulin. This should be regarded as a very decisive metabolic trait. Froesch et al. [283, 284, 291, 564] have based their arguments on experiments in which rats were treated with anti-insulin serum or in which experimental diabetes was induced in animals with streptozotocin. Thus other findings obtained with human subjects should entitle to the ideas presented by Lang [289]. Furthermore, the basis for the scruting should be the caloric utilization of xylitol. This phase of the utilization, the transfer through the plasma membrane, is insulin-independent. Table 8 later lists additional studies related to the involvement of insulin in xylitol metabolism. Furthermore, other animal experiments with isolated liver preparations as well as infusion experiments in humans [285, 292] do not seem to support the idea of the formation of high amounts of lactate from xylitol [292], although xylitol administration increased the lactate-pyruvate ratio.

Intravenous infusion of sorbitol and xylitol was shown to lead to an increase in serum urate concentration [292, 293]. Feeding studies for 2 weeks [500] and for 2 years (Turku sugar studies) excluded, however, that inclusion of xylitol into a normal diet would increase urine or serum urate concentration [294]. Such an increase could only be observed in serum under special metabolic conditions, e.g. after a sucrose-free preexperimental period [295–298, 602]. Therefore, this temporary increase has to be regarded as a normal response of the body to high oral doses of sugars and not as an adverse reaction, because sucrose induces a similar effect [295, 296, 602].

Förster [573] discussed the possible side effects of oral administration of glucose, fructose, sorbitol and xylitol in man. He concluded that there are no specific and important side effects of these three glucose substitutes, as compared to glucose. According to Förster [573] there are no facts that would support the hepatotoxicity of xylitol. The danger of hyperglycemia was considered to be smaller with xylitol, sorbitol and fructose as compared to glucose.

On parenteral administration hyperuricemia is caused by fructose, sorbitol and xylitol, but not by glucose or galactose [540]. A highly-dosed and rapid infusion (for example, 1.5 g/kg in 20 min) increased the concentration of serum urate. Simultaneously the renal excretion of urate was increased. Förster has further stated [573] that during continuous infusion the effect of xylitol on serum urate concentration was very small. After 48 hours of xylitol infusion at a dose of 0.25 g/kg an increase of 1–2 mg/100 ml in the concentration of serum urate was seen [573]. Sorbitol at the same dose was without any effect.

Förster then draws the attention to the following three points: (1) Distinct differences exist between the results of animal experiments and studies in humans. (2) It is possible that different mechanisms contribute to the increase in purine synthesis by various carbohydrates. (3) The elevation of serum urate concentration is so small that in most cases it can be neglected [573].

There is, however, another side effect of xylitol infusions which may be encountered if attention is not drawn to the mineral metabolism of the patients. During the first hours (12 hours in an experiment of Förster [573]) of xylitol infusion the renal excretion of potassium was almost twice as much as during glucose infusion. Sorbitol was less effective than xylitol. However, after prolonged infusion of xylitol, the potassium excretion was decreased to normal values. This finding is similar to that made by Berg et al. [574] who reported that a 10–12 hours' infusion of xylitol caused nausea and vomiting. An application of 50 meq potassium per day prevented this type of complications [573]. Also Heller [575] reported side effects of infusions with xylitol. The effects were considered to be due to the loss of potassium ions. When potassium ions were substituted, the symptoms disappeared. It should be realized that all nutritional carbohydrates react in their typical way. Attention should thus be drawn to sufficient addition of necessary compounds during infusion of each type of carbohydrate.

Claims have been occasionally presented that intravenous administration of xylitol would also increase serum bilirubin concentration [264, 265, 298, 299] and alkaline phosphatase activity [299]. A part of these experiments were carried out on rats [299] and in the other experiments [264, 265, 298] the subjects were loaded with fairly high quantities of xylitol within a shorter period. Meng [300] observed that intravenous infusion of xylitol did elevate the activity of serum alanine aminotransferase and alkaline phosphatase in dogs. However, these changes were dose-dependent and reversible. Moreover, glucose in parenteral nutrition also produced an increase in the activity of these enzymes [300]. Xylitol administration did not produce significant changes in sulfobromphtalein retention and serum bilirubin [300]. As regards the above parameters sorbitol seemed to behave in the same way in dogs as xylitol. Meng [300] suggested that simultaneously given amino acids may have contributed to the above alterations. Förster et al. [293] showed with healthy human subjects a small increase in serum bilirubin concentration, but the increase was of the same order as that obtained with glucose, sorbitol or fructose. 24 hours later the values had returned to the initial level. In this context Förster [292] recalled the fact that an increase in serum bilirubin following intravenous glucose infusion was reported long ago by Popper and Schaffner [301]. The regularity of this type of biochemical responses should be emphasized.

In the approximately fifty volunteers who were fed a xylitol-containing diet for 2 years in the Turku sugar studies, no increase in serum alkaline phosphatase and transaminase activity or bilirubin concentration was observed [294, 302]. Similar results were obtained in rat experiments in which animals were given glucose, fructose, sorbitol and xylitol by intravenous infusions [293]. Transaminases and bilirubin were in all instances lower than in control animals. No hepatotoxicity as a result of the use of these carbohydrates was found.

In special animal experiments Förster [292] clarified that the effect of sugars and sugar alcohols on renal urea excretion is an extra-renal one. The nitrogen-sparing effect of xylitol and fructose was also suggested to be a direct one, and not mediated by the primary conversion to glucose of these sugars. Fructose and xylitol infusions did not have any effect on urea excretion on the first infusion day of a 72-hour experiment [292], but these compounds were shown to prevent the increase in urea excretion on the following days. In spite of high infusion rate [0.7 g/(kg$\times$hr)], the urinary excretion of these sugars was low. For sorbitol and xylitol it corresponded to 20% of the intravenous dose. 22-month chronic intake of sucrose, fructose or xylitol did not affect serum urea concentration to any noticeable extent [302].

Xylitol infusion has been shown to reduce the concentration of free fatty acids in human volunteers [292]. In high dose xylitol stimulates insulin secretion in man. When xylitol is largely metabolized by the liver, there is said to be a stimulated hepatic esterification of circulating free fatty acids and a diminished peripheral release of fatty acids [292], a combined mechanism leading to the reduction of free fatty acids.

7.24 Base excess in infusion therapy

Bickel [557] discussed the possible reasons for the observed decrease of base excess after the infusion of xylitol. For this discussion it is necessary to recall that an infusion of glucose, fructose or a carbohydrate mixture of glucose, fructose and xylitol, also decreases the quantity of basic equivalents. This may be largely due to the increase of acid equivalents in the form of lactate, and to a certain extent to a dilution, as the water balances are positive [557]. Virtually all studies indicate, however, that with the infusion of xylitol at a moderate rate, the accumulation of lactate is very small. According to Bickel [557] this low formation of lactate does not at all explain the decrease of the base excess. Bickel [557] provided the following three explanations:

- Due to an osmotic overload, an increased fluid shift from the intracellular to the extracellular space occurs. This would result in a further decrease in the extracellular buffering capacity.
- In addition to lactate, still other metabolic acids may accumulate.
- It is known that xylitol does not undergo renal reabsorption (fructose and glucose are reabsorbed by the kidney) [388, 583]. It has been shown that substances that are not reabsorbed by the kidney, or which are absorbed only to a minor degree (for example, mannitol and urea), cause an intensified excretion of HCO_3^- ions [584, 585]. The resulting bicarbonate loss, persisting over a long period, may lead to a decrease of the basic equivalents in the blood.

When Bickel provided these explanations, sufficient data were not available about the excretion of HCO_3^- ions during infusion of xylitol in man. An interesting point of comparison can be found in dentistry. It is customary that dentists and physiologists do not usually realize the natural and mutual connection of their scientific fields. Although the oral cavity forms a separate entity which cannot always be compared to the internal systems of man, comparisons of data may often be valuable. For example, when human volunteers chewed chewing gums sweetened with either sucrose, fructose, sorbitol or xylitol, there was a clear difference between the resulting concentrations of HCO_3^- ions in whole saliva immediately after a 2-minute stimulation period [568]. The excess of these ions was most likely derived either from the main or minor salivary glands or directly from the blood through the oral mucosa. The loss of the glandular or blood bicarbonate was naturally of very small quantitative importance as each piece of chewing gum contained only 1.5 g of the sweetener involved (fig. 6).

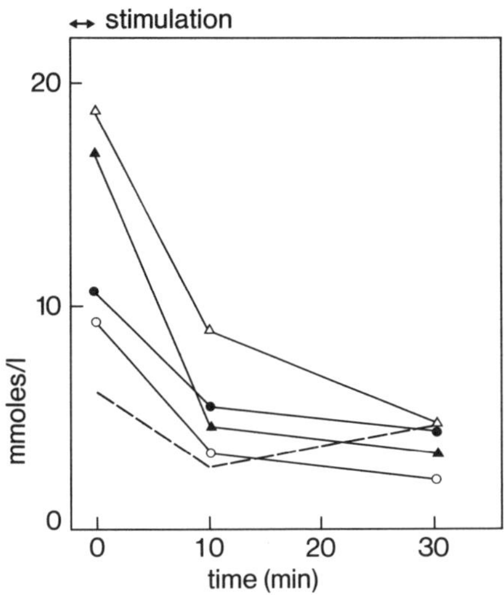

Figure 6
The concentration of HCO_3^- ions in human whole saliva after stimulating the salivary flow rate with sweetened and unsweetened stimulators. Arithmetic means of ten adult subjects. Stimulation time was 2 minutes. Stimulators: ●, chewing gum base; ----, paraffin; △, xylitol chewing gum; ○, sucrose chewing gum; ▲, sorbitol chewing gum. From [568].

7.25 Aspects of the 2-year Turku feeding study

General and clinicochemical description of the Turku sugar studies has been provided separately [133, 294, 302] and a number of details have already been touched upon in this review. In this study the dental and medical effects of dietary sucrose, fructose and xylitol were compared. As a conclusion the following findings will be emphasized:

(a) Most subjects maintained excellent cooperation and adhered to the dietary regimen for 2 years. This was primarily due to the availability of a wide assortment of xylitol-containing foodstuffs and an effective distribution organization. A decisive factor seemed to be the good quality of virtually all xylitol and fructose products.

(b) Long-term intake of xylitol, fructose or sucrose did not cause clinically significant differences between the sugar groups in any of the following serum or blood parameters:

- Hemoglobin, red blood cells, leukocytes, sedimentation rate of red cells.

- Pyruvate, lactate, urate.
- Triglycerides, cholesterol, glucose.
- Insulin.
- Bilirubin.
- Aspartate and alanine aminotransferase.
- Alkaline phosphatase.
- Lactate dehydrogenase.
- Amino acids (including certain other ninhydrin-positive compounds). (See subsequent chapters regarding certain amino acids.)
- Na, K, Ca, Mg, inorganic phosphorus.
- Amylase (see [302] for an explanation of normal differences between groups).
- Immunoglobulins A, G and M.
- Ascorbic acid.
- Disc-electrophoretic [303] and isoelectric focusing pattern of serum proteins.

(c) Urine urate concentration was not changed as a result of chronic intake of xylitol or fructose [294].

(d) Average monthly intakes of sucrose, fructose and xylitol were 2.2, 2.1 and 1.5 kg, respectively. In numerous cases the daily xylitol doses exceeded 100–200 g. The differences between the test groups were not due to the quality of the xylitol products. As previously mentioned, virtually all foodstuffs made of xylitol were very acceptable. There were several understandable reasons to the above differences in consumption:
   - When the nutritional trial was commenced, xylitol was a new-comer in food industry. It was not possible to create in 1 or 2 years as versatile assortment of xylitol products as it has been possible to do in over 100 years using sucrose.
   - In a few subjects high doses of xylitol produced transient osmotic diarrhea. This could be avoided by controlling the quantities of xylitol consumed.
   - The subjects received free of charge only those sucrose and fructose products from the distribution centre, which could be made of xylitol. The subjects of course obtained certain amounts of other sucrose and fructose produce (the latter are available in abundance in Finland from normal food stores) through normal routes.

   These matters were also discussed in report No. VI of the Turku sugar studies (Acta Odont. Scand. *33*, Suppl. 70, 1975).

(e) Osmotic diarrhea which occurred in several xylitol-consuming subjects during the first weeks of the trial, disappeared almost totally after an adaptive period. The length of this period was individual and it varied from 3 days to 3 weeks. More than half of the approximately fifty xylitol-consuming subjects did not report diarrhea or flatulence at any stage. During later phases of the trial these disorders were reported in the sucrose and fructose groups almost as frequently as in the xylitol group. When xylitol subjects were switched to sucrose diet after 2 years, this change resulted in diarrhea in many test persons.

(f) Human subjects have an individual ability to tolerate xylitol, a fact which concerns the intake of most ingredients of human diet. A few subjects (5 out of 53) were more sensitive, reporting stomach disorders, more or less subjectively, in connection of daily doses of 20–30 g xylitol. The majority reported no disorders with daily doses of 50–100 g. As emphasized in other connections, several subjects exceeded the 100–200 g/day dosage without any side effects. The highest calculated xylitol intake was 430 g/day without any such side effects which would differ from those exerted by similar amounts of sucrose or fructose.

(g) During the course of the trial several pregnancies appeared. No abnormality was observed in any of the pregnancies, deliveries and infants in the three test groups. Table 5 lists certain serum parameters of pregnant sucrose- and xylitol-consuming subjects. All values fell within the normal clinical range. During the course of five pregnancies which occurred in the xylitol group, the monthly xylitol consumption was 0.8–1.0 kg in four cases and 0.5 kg in one case. Thus pregnant subjects used less xylitol than other test persons. A part of this lower consumption resulted from normal restrictions during the periods of hospitalization. (Females in general used in most cases less sugars than males, but there were a few pronounced deviations from this rule [133].) At the time when this review was submitted for publication (April 1977) no abnormalities have occurred in the babies or in the parents. No teratological findings have been made. The study was carried out during 1972–1974.

(h) One diabetic (female, aged 33 years) was on strict xylitol diet for 2 years. No adverse effects were ever encountered. Blood and urine chemistry values became more constant during xylitol consumption than what was found prior to the start of the xylitol regimen. Table 6 lists certain parameters of this subject. After 4.5 years of the commencement of the trial (when this review was completed), the subject was in good health. She has continuously used xylitol after the termination of the 2-year trial. The clinical condition has remained unchanged.

(i) A number of the Turku study subjects have continued the consumption of various xylitol products after the termination of the trial. After the end of the study the consumption of xylitol has naturally been lower than during its course. The present experience, after almost 4,5-year consumption of xylitol, indicates that in children aged 3–17 years no side effects have been observed with a daily intake varying from 3 to 50 g. Osmotic diarrhea and flatulence have not occurred [614].

(j) It was not possible to thoroughly trace the serum amino acid composition of all Turku study subjects. Hence no very firm conclusions can be drawn. However, it is interesting to notice that serum glutamine-asparagine values were almost constantly lowest in the xylitol group [302]. Similarly, serum glutamate values were almost always highest in xylitol-consuming subjects. However, all amino acid values were within the normal clinical range.

(k) The consumption of a sucrose diet produced a higher serum amylase activity than the consumption of fructose or xylitol diets. This is an expected and normal response of pancreas.

Table 5
Chemical and hematological values of pregnant sucrose-, fructose- and xylitol-consuming subjects of the Turku studies [294, 302].

Subject No.	2	7	8	8¹)	22	32	36	39	51	52	72	74	82	105	105¹)	106	110
Month of pregnancy²)	IV	V	IX	IX	II	VIII	VI	VII	III	V	VIII	IX	IV	VII	VI	III	IV
Sugar group	F	S	X	X	X	F	F	S	S	F	S	X	S	X	X	F	S
Sugar intake³)	2.8	2.8	1.0	1.0	0.8	1.5	1.2	1.5	1.1	1.6	1.0	0.9	1.7	0.4	0.5	1.0	1.0
Na (mmol/l)		136.7		123.9	124.3				130.4		135.0				118.3		
K (mmol/l)		4.15		3.96	3.71				3.93		4.22				4.21		
Ca (mmol/l)		2.16		2.31	2.30				2.49		2.33				2.12		
Mg (mmol/l)		0.79		0.81	0.79				0.79		0.66				0.68		
P_i (mmol/l)		0.73		0.84	0.94				1.09		1.03				0.96		
Alkaline phosphatase⁴)		1.09		3.51	0.56				1.29		0.95				1.04		
Amylase (U/l)		80.0		32.1	38.0				25.0		21.6				18.3		
IgA (g/l)		2.20		1.89	0.65				2.52		2.33				0.96		
IgG (g/l)		13.75		9.24	10.56				10.01		13.09				9.13		
IgM (g/l)		2.25		1.48	1.38				2.17		2.11				0.99		
Alanine aminotransferase (U/l)	16.3	5.5	19.4	9.4	3.3	4.3	7.4	11.1	5.5	9.1	12.2	7.6	2.7	9.6	7.5	2.4	1.4
Aspartate aminotransferase (U/l)	13.3	9.7	13.6	14.2	7.9	7.8	6.7	8.5	8.2	12.0	9.4	8.1	8.0	5.0	7.0	7.95	5.9
Lactate dehydrogenase (U/l)	125	288	240	250	192	144	288	192	154	134	163	250	365	288	317	220	288
Hemoglobin (g/l)	112	105	133	117	113	134	116	131	122	112	126	132	140	108	108	126	118
Leukocytes ($\times 10^6$ liter^{-1})	10,260	6634	12,460	6434	6834	5566	10,434	9166	7200	6900	9966	5534	9300	6134	5366	9166	8100
Sedimentation of red blood cells (mm/hr)	41	49	41	47	20	40	19	47	9	30	22	35	11	40	47	25	28
Glucose (mmol/l)	5.1	4.4	3.9	4.7	4.9	4.0	3.6	4.1	4.0	3.8	5.8	4.7	4.3	4.9	4.3	4.1	4.2
Triglycerides (mmol/l)	1.58	1.54	2.0	2.21	0.62	2.45	1.56	2.25	0.58	1.60	3.26	2.45	1.09	1.23	1.31	0.80	0.87
Urate (serum) (mmol/l)	0.24	0.20	0.18	0.25	0.21	0.26	0.28	0.24	0.25	0.22	0.21	0.35	0.27	0.21	0.20	0.25	0.21
Urate (urine) (mmol/24 hr)	5.18	3.75	4.75	4.20		4.06	5.40	4.35		4.36	4.50	3.72		3.39	2.26		
Cholesterol (mmol/l)	6.1	6.0	6.6	7.2	4.9	6.9	6.5	6.8	3.8	5.9	7.8	8.4	6.5	7.1	7.7	5.0	4.9

1) Second pregnancy.
2) Month of pregnancy at the time of blood collection.
3) Average monthly use of sucrose, fructose or xylitol during pregnancy until blood collection (in kg).
4) Bessey-Lowry units.

Table 6
Chemical and hematological values of a diabetic subject given xylitol enterally during a period of 2 years. The subject (female, aged 30 years at the end of the study in 1974) received 2.2 ml insulin per day. The disease was detected in 1961. Consumption of xylitol: 65 g/day[1]).

Parameter[2])	Time (Months)					
	0[3])	5.5[3])	10.5[4])	14[4])	18.5[4])	22[3])
Hemoglobin (g/l)		122	140		131	136
Leukocytes (in mm^3)		7060	3766		4966	8034
Sedimentation (mm/hr)		15	6		30	8
Alanine aminotransferase (U/l)		9.1	0.9		4.3	3.5
Aspartate aminotransferase (U/l)		10.5	4.8		7.8	6.3
Lactate dehydrogenase (U/l)		192	297		269	288
Lactate (mmol/l)		1.17	1.67		1.77	1.60
Pyruvate (mmol/l)		0.031	0.068		0.141	0.093
Cholesterol (mmol/l)	4.7	4.3	5.9	4.6	4.9	5.1
Triglycerides (mmol/l)	1.24	1.71	1.17	0.99	1.30	1.15
Glucose (mmol/l)	12.8	4.0	17.5	6.0	6.6	19.9
Urate (mmol/l)	0.18	0.22	0.21	0.26	0.29	0.21
Urate, urine (mmol/l)	2.65	1.66			2.65	

1) At the 22-month phase the following parameters were also determined: Amylase, alkaline phosphatase, Na, K, Ca, Mg, P_i, IgA, IgG, IgM, insulin. All values were normal.
2) Serum parameters (urate was determined in both serum and urine).
3) Non-fasting state.
4) Fasting state.

(l) The chronic 2-year consumption of the xylitol diet did not impair the absorption of iron, as deduced from the hematological data [302].

(m) As numerous subjects have continuously been engaged in the activities of the Dental School, it has been possible to observe their general health condition. No information about cataract or other pathological changes has been obtained after 4.5 years since the start of the 2-year feeding study.

7.26 Xylitol in erythrocyte metabolism

The metabolic capacity of red blood cells is largely directed toward the maintenance of the cell membrane and hemoglobin molecule. The intermediary metabolism of the cells is mainly limited to glycolysis as the energy-producing pathway, yielding 2,3-diphosphoglycerate which is required for normal hemoglobin functions. Red blood cells also possess pentose phosphate shunt which is likewise needed in the maintenance of the cellular integrity.

The shunt produces NADPH which in the erythrocytes is largely used to keep glutathione in the reduced state. An unnecessary high rate of glutathione oxidation may lead to cell injury. The glutathione system is thus partly responsible for the integrity of the cell membrane. Accumulating excessive oxidized glutathione will leave the cell and new reduced glutathione is synthesized in the cell [279, 304, 305].

Agents or conditions which induce changes in the glutathione metabolism in red cells may lead to hemolysis. The ultimate reasons for red cell damage can be nutritionally [306] or genetically determined. For example, inability of the red cells to provide glucose 6-phosphate dehydrogenase for the reaction glucose 6-phosphate →6-phosphogluconic acid, is encountered in 3% of the world population [307]. As a result of this enzyme deficiency, there is a limited supply of cellular NADPH which is normally generated in dehydrogenations of the sugar phosphate. Sufficient NADPH is needed by glutathione reductase to keep an adequate amount of glutathione in the reduced state. There are several genetic variants of this enzyme deficiency and, accordingly, several clinical syndroms have been recognized, including drug-induced hemolytic anemias, neonatal anemia and secondary chronic hemolysis [279, 308, 309].

Xylitol has been regarded as a possible therapeutic agent in certain type of anemias because it potentially produces NADPH through its oxidation to the ketopentose (L-xylulose). For this purpose it is naturally required that erythrocytes would metabolize xylitol. The works of Asakura, Yoshikawa, van Eys, Wang and others [277, 311–313] showed rapid uptake of xylitol by red blood cells and suggested the presence of xylitol dehydrogenase activity in these cells. Wang and van Eys [279, 314] showed that xylitol, sorbitol, ribitol, D-mannitol, galactitol, L-arabitol and D-arabitol were metabolized in the erythrocytes with NAD as coenzyme. Glycerol, D-erythritol or inositol were not oxidized. Only xylitol was metabolized to a noticeable extent when NADP was used. Sorbitol was oxidized slowly. It was also shown that intact erythrocytes are able to utilize xylitol [279]. Xylitol addition to blood stored in a citrate-glucose solution resulted in the regeneration of ATP and other high-energy phosphates, xylitol and inositol being additive in this effect [279, 311]. It is possible that xylitol in general induces the generation of new nucleotides via de novo biosynthesis of purines. Infusion of xylitol at a rate of 0.4 g/(kg$\times$hr) increased the generation of 5-phosphoribosylpyrophosphate which is an important substrate in the biosynthesis of purines [587].

Xylitol was shown to have a beneficial effect on the osmotic fragility of stored blood [311]. Xylitol was found to affect the level of glycolytic intermediates through changes in the ratio of reduced and oxidized nicotinamide coenzymes [277]. The production of lactate in erythrocytes in the presence of xylitol [277, 316] has been explained by the change of $NADH_2/NAD$ ratio in the cells [277]. Thus there would not necessarily be any utilization of xylitol to lactate.

The group of Asakura [312] also showed that the rate of reduction of methemoglobin in fresh erythrocytes in a 50–500 mM glucose medium was constant whereas in a xylitol medium the reduction was concentration-dependent. Stored erythrocytes barely reduced methemoglobin, whereas upon addition of xylitol they reduced methemoglobin well [315].

Van Eys et al. [279] concluded that three syndromes in glucose 6-phosphate dehydrogenase deficiency require therapy: favism and other extreme drug-induced hemolyses, neonatal jaundice and congenital non-spherocytic anemia. It is expected that new variants of these diseases will be discovered making the availability of an effective therapy necessary. So, for example, variants associated with chronic non-spherocytic hemolytic anemia, or with drug-induced hemolysis, have been described in geographical areas in which glucose 6-phosphate deficiency is normally rare [317–319]. There is no reason to believe that the incidence (or rather discovery) of such anemias would be decreasing. Because orally administered xylitol is very well tolerated by man, the potential of xylitol in rational therapy of these diseases should be studied. A single dose of xylitol (0.5 g/kg) did not affect the activity of catalase and glucose 6-phosphate dehydrogenase of erythrocytes [603].

7.27 Effect of xylitol on exocrine gland function

The consumption of a xylitol diet influences in a dose-dependent manner the enzyme, glycoprotein, mucopolysaccharide and electrolyte secretion of human and monkey salivary glands. Salivary glands in general respond to the quantity and nature of the carbohydrate portion of diet. These effects were also dealt with in the dental part of this review. So, for example, xylitol consumption increased the concentration of inorganic phosphate and pH [568], and the activity of lactoperoxidase [223] in human whole saliva when compared to sucrose consumption. Lactoperoxidase [206] and α-amylase [245] activities were also increased in submandibular and parotid saliva of xylitol-fed monkeys *(Macaca mulatta)* when compared to sucrose consumption. The concentration of total protein was simultaneously elevated [206, 245], but xylitol administration most likely induced the secretion of certain specific glycoproteins as well. Gastric intubation of 2.5 g xylitol per day for 3 days did not significantly affect parotid saliva lactoperoxidase activity [244] in monkeys *(Macaca fascicularis)* when compared to intubation of sorbitol.

The ability of xylitol to produce the above-mentioned effects is not unique. Sucrose, glucose, sorbitol and other sweet dietary carbohydrates also increase salivary Ca, phosphate (particularly HPO_4^{2-} ions) and pH in man, as well as the flow rate of saliva. Depending on a number of factors, a concomitant pH drop may result in oral fluid after intake of fermentable carbohydrates. Such factors include, for example, the state of oral hygiene and buffering capacity of secretions. Particularly sucrose may finally lead to lowered pH values, whereas in the presence of sufficient amounts of xylitol no pathological pH drop is encountered in man.

In addition to these common effects dietary carbohydrates also exhibit selectivity of a certain degree in their effects upon human saliva. While salivary lactoperoxidase [90, 91, 206] or α-amylase [245] were increased on xylitol diet, no such effect has so far been found with salivary IgA, IgG or IgM in man [91]. Such differences between the effects of various dietary carbohydrates should be regarded as normal, and not as pathological changes. On the contrary, it is believed that the increased salivary lactoperoxidase activity on xylitol diet, found in man [223] and monkeys [206, 244], has several dentally advantageous consequences, as in the fight against pathogenic oral microorganisms. As to the lactoperoxidase effect, there may not be pronounced differences between xylitol and sorbitol [244], but this matter should be more adequately elucidated.

It is possible that all mucous membranes of the alimentary tract behave more or less similarly in the presence of polyols. The effects are perhaps more clearly seen in the oral cavity, because ingested polyols are effectively diluted in other parts of the tract. It is also possible that a part of the changes in the concentration of inorganic ions in human oral fluid after xylitol intake is due to rapid flux of ions and water through oral mucosa, and not solely due to responses of the main three pairs of salivary glands. Anyway, as the consumption of polyols seems to affect the salivary glands and pancreas, the response of other exocrine glands should also be considered.

Oral administration of moderate quantities of xylitol has a more direct and stronger effect on the production of hormones from gastrointestinal exocrine glands than from endocrine glands, although glucagon and insulin production from pancreas may be influenced by xylitol to a noticeable extent depending on dose, animal species and experimental conditions. Amylase production may also be slightly affected [302]. The gastrointestinal hormones which evoke interest, are listed below:

- Gastrin I and II (in pylorus mucosa, stimulate the secretion of HCl in the stomach, possibly also with the aid of histamine).
- Secretin (in the intestinal mucosa, stimulates the pancreatic and biliary secretion of water and HCO_3^- ions, prevents stomach HCl secretion, stimulates insulin secretion).
- Pancreozymin (in the intestinal mucosa, stimulates pancreas amylase, trypsinogen and lipase secretion).
- Cholecystokinin (in the intestinal mucosa, stimulates the emptying of the gallbladder).

No certain information about the effect of xylitol on the secretion of these hormones is available. It is likewise not known with certainty whether dietary sugars have a selective effect on the function of exocrine glands other than the salivary glands. However, it has not yet been possible to show any clear correlation between xylitol intake and the activity of lactoperoxidase in monkey [206, 244] or bovine [604] lacrimal fluid.

Recent studies [604] suggest that it may be possible to influence the enzymic properties of exocrine secretions through the carbohydrate composition of diet. The milk of cows fed a polyol mixture (chiefly mannitol, xylitol, sorbitol and arabinitol) had a more constant and higher lactoperoxidase activity than milk of the control groups. The possibility that the higher lactoperoxidase activity would be associated with improved preservation qualities of the milk, is being elucidated. The use of polyol mixtures as fodder in the feeding of certain domestic animals may have positive effects on the weight gain and mortality of offspring. These effects could be achieved by feeding the dams with polyols during the lactation period.

7.28 Polyol cataracts

Orally administered xylitol and sorbitol have failed to cause cataract formation in human subjects and experimental animals. Certain polyalcohols are still involved in sugar cataract. The reason why oral administration of xylitol does not induce cataract in healthy human and animal subjects habitually consuming this carbohydrate, may thus deserve a closer examination.

A cloudiness or opacity in the lens of the eye, which interferes with vision is called cataract. Polyol or sugar cataracts are formed, for example, in diabetic test animals as a result of accumulation of polyols formed within the lens cells. In most cases the cause of a cataract can not be readily shown, although changes in the chemical composition of the lens are known to occur. Cataract may result from normal ageing of the eye (senile cataract), infection, mechanical injury, chemical and genetic factors, etc. [320, 321]. Animal experiments seem to indicate that multiple cataractogenic factors have additive and synergistic effects [322–324].

Osmotic swelling of the lens seems to be a common feature in many experimental cataracts, including sugar cataract [320, 321]. Although the nature of the initiating factor may vary, the final stages often include an influx of Na^+ and Cl^- ions. The normal state of hydration and the lens volume are balanced by permeability of the lens membrane and by the efficacy of a cation pump [321]. Na^+ ions are extruded and K^+ ions are taken in. The pump mechanism partly prevents a Donnan type of hydration.

Cataractogenic sugars (glucose, galactose and xylose) are associated with lens swelling. An initiating factor is the elevation of aldose concentration in the aqueous humour. This in turn increases the concentration of the corresponding polyol in lens. For example, in vitro incubation of rabbit lens in 30 mM galactose or in ouabain results in immediate lens swelling [321]. The cataractogenic sugars are converted in lens to sugar alcohols by aldose reductase [325]. The formed sugar alcohols are not metabolized effectively any further. Because they are neither translocated across the lens membranes, there is a concentration of polyols in lens fibers. The resulting hypertonicity induces an inflow of water.

Galactose cataracts in rats can be easily induced by feeding the animals diets enriched with galactose. If rats are placed on a high galactose diet, opacities occur in the equatorial region on the third day and become progressively more severe until the formation of a non-reversible mature cataract in 2 weeks [326]. Galactitol which is formed in the lens cells, accumulates and causes an osmotic swelling [326–330]. The level of ATP is markedly depressed in the later stages of cataract development [320]. Consequently, lack of energy may be a contributing factor in later changes. The change in hydration is also usually accompanied by loss of amino acids in the lens when exposed to galactose. This is most likely primarily due to an osmotic swelling of the lens brought about by the retention of galactitol [320]. There is also a dilution of cations in the initial vacuolar stage [331].

The most active substrate for lens aldose reductase is xylose. Galactose is a fairly good substrate and glucose is least active [332]. The enzyme triggers the mechanism which leads to sugar cataract [321]. Enzyme inhibitors of aldose reductase (for example, 3,3-tetramethylene glutarate) cause a delay in the cataractous process or prevent it.

In the early stages of sugar cataract formation the process is reversible, but when the lens is maintained in the swollen state for a prolonged period, there is a loss of amino acids and redistribution of cations [334]. In vitro incubation experiments with lenses have generally shown that the initial osmotic change parallels the accumulation of polyols, for example, sorbitol. Later stages are characterized by a decrease of the concentration of sorbitol. Then the increase in hydration becomes more dependent on electrolyte changes between the lens cells and their environment [321].

When a diet containing 35% of D-xylose was fed to a weaning rat, complete opacity of the lens ensued [335]. This phenomenon is based on the fact that xylose penetrates readily from the blood into the aqueous humour [336]. The lens contains an enzyme which normally converts xylitol to L-xylose. The equilibrium is in favour of formation of xylitol, but continuous removal of D-xylose forces the reaction in the direction of breakdown of xylitol. However, the penetration of D-xylose into the aqueous humour of the rat lens led to an accumulation of both xylitol and sorbitol in the lens. Sorbitol was not detected in the lens of glucose-fed controls. Xylitol did not accumulate in other tissues of the xylose-fed rat, except in the aqueous humour where a small amount was found. The first signs of cataract usually appeared after 5 days. No xylulose, xylonic acid or xylose phosphates were detected in lens. Very little xylose was present [336]. No differences were observed in lactate and various phosphate fractions between the lens of xylose-fed rats and in those of glucose-fed controls.

Based on these literature citations, the knowledge about rat and rabbit sugar cataracts can be said to be fairly comprehensive, but the situation in man is less clear. It is known that sorbitol, glucose and fructose accumulate in senile cataractous lenses of diabetics removed at operation compared to either non-cataractous lenses removed at post-mortem or to senile cataractous lenses removed from non-diabetics [337]. In a recent study, Heaf and Galton [338] confirmed that although levels of glucose were significantly elevated in lenses of diabetics, sorbitol was always detected in lenses of diabetic and non-diabetic patients and no significant differences were noted. The discrepancy between these two studies [337, 338] was explained by the use of a less sensitive sugar and polyol assay in the earlier study [337] (paper chromatography) than in the later one [338] (gas-liquid chromatography). On the other hand, both methods yielded similar amounts for inositol in lenses. Paper chromatography is able to reveal inositol more reliable due to the high levels of this polyol in lenses (4.5 mg/g) [337, 338]. Another gas-chromatographic study of polyols in lenses revealed accumulation of galactitol and sorbitol in the lens of an infant with galactosemia, but no polyols were found with this sensitive method in the lens of three control infants dying, respectively, from sepsis, prematurity and congenital heart disease [339].

Thus it seems possible that tissue accumulation of polyols could be responsible for the secondary complications of diabetes. Surveys made thus far indicate that the incidence of cataracts in diabetic subjects is not different from that in non-diabetic subjects [321]. However, other surveys made in Great Britain [340] and Germany [341, 342] showed that the frequency of cataract extraction is much higher in diabetic than in non-diabetic subjects. This has been claimed to indicate that the maturation of cataracts in diabetics occurs sooner than in non-diabetic subjects. As pointed out by Kinoshita [321], true diabetic cataracts are uncommon.

When elucidating the mechanism of xylose cataract, galactosemic cataract provides an interesting point of comparison. Galactosemia is a hereditary disease which is recessive in nature. Its occurrence is approximately 1 in 18,000 births [343]. The disease is caused by a deficiency of one enzyme, most likely galactose 1-phosphate uridyltransferase, needed in the metabolism of galactose. Human galactosemic patients [344] and galactosemic rats [345] have an amino aciduria. Galactitol can be found in the urine of human galactosemic patients [346] and in the kidney of galactose-fed rats [320, 346]. Thus the same factors may affect both in lens and other tissues in galactosemic subjects.

Lens is a particularly suitable site for the accumulation and production of galactitol. Four main enzymic properties are responsible for the differences found between the fate of various polyols in lens. These are as follows:

(1) Based on the kinetic properties and affinity characteristics of lens aldose reductase, it appears that galactose concentration must be fairly high before the enzyme is able to convert significant amounts of the aldose to the polyol form. This is nature's way of protecting the lens from unnecessary rapid polyol formation in lens cells. In other human tissues, even though the organ would be exposed to high levels of galactose, the activity of galactokinase is sufficiently high to keep the sugar concentration at low levels. The phosphorylation mechanism of galactose in lens acts at a fairly low rate and there is a tendency for the sugar to increase when it is available in higher levels to this tissue.

(2) The lens cells have, however, an active mechanism which provides NADPH, needed in the reduction of galactose.
(3) Lens aldose reductase is relatively active.
(4) In contrast to other sugar alcohols, galactitol is not a suitable substrate for polyol dehydrogenase, an enzyme responsible for the further oxidation of polyols.

Most tissues actively oxidize sorbitol and xylitol, but not galactitol, to their keto forms. Because sugar alcohols do not easily penetrate the lens membrane, galactitol does not leak out of the lens. Because it is not rapidly metabolized either, its continued synthesis may lead to its accumulation.

Many dietetic properties and physiological reactions of sorbitol, xylitol and fructose are the same as those of glucose. All three first mentioned sugars are partly converted into glucose in various tissues. A main difference between the metabolic reactions of these sugars is that glucose is metabolized largely by peripheral tissues in the presence of insulin, whereas sorbitol, xylitol and fructose are to a high extent treated by liver independent of insulin. Even during prolonged and heavy loading of human volunteers with oral xylitol, this polyol is absorbed at a suitable rate and transported into the liver cells where perhaps more than 80% of ingested xylitol is treated. The main product is glucose which is subsequently released to the circulation [284, 292]. Orally or intravenously administered xylitol is not transported into lens cells, but is converted into glucose almost exclusively in the liver. There is thus a natural explanation why orally administered sorbitol and xylitol do not induce cataract in healthy subjects. The recent Turku sugar studies, involving 2-year chronic intake of high amounts of xylitol in approximately fifty non-diabetic and one diabetic subjects, showed that the connection between xylitol intake and cataract was undemonstrable. This has earlier been shown in numerous animal and human experiments. Furthermore, even the xylose-associated cataract in experimental animals has not been adequately established; rats seem to develop it after heavy loading with xylose, but these results should not directly be applied to domestic animals, like the ruminants.

7.3 **Safety of xylitol**

7.31 The Australian cases

Preceding chapters have provided literature quotations which strongly suggest that man tolerates xylitol very well and that many clinicochemical effects of xylitol in man are also produced by glucose, fructose and sucrose. Against this background it was interesting, but simultaneously upsetting that certain adverse reactions were encountered in parenteral feeding with xylitol in Australia [263, 265, 347, 348] and in the U.S.A. [349]. While this type of findings are highly significant, per se, and while they emphasize the importance of alertness in the use of any substance in parenteral nutrition, they nevertheless have also led to unnecessary and confusing consequences among health authorities. The history of the Australian cases has been largely clarified.

In Australia, xylitol was used for parenteral nutrition on a large scale without any communications being published on any side effects. Similar situations prevailed in Japan, West Germany and other countries. At those times, up to 1969, most xylitol used for parenteral feeding was obtained from Europe. Later fairly large quantities of xylitol were being manufactured for diabetic subjects in the Soviet Union.

Certain Japanese inventions led to a small-scale industrial xylitol production in this country. In 1969 and 1970 twenty patients who received Japanese xylitol in South Australia died displaying symptoms of acidosis, liver necrosis and renal oxalosis. Crystals of calcium oxalate were identified in tissues from five patients given intravenous solutions containing xylitol [349]. Health authorities including the Australian Drug Committee and the WHO were immediately informed. Xylitol was recommended to be withdrawn from clinical use and prohibited as an import into this country [348].

An Australian patient with very severe liver damage received only approximately 200 g of xylitol. On the other hand, in no single case in which higher quantities of xylitol were given in Japan and West Germany, similar reactions were observed. In two other studies in which xylitol was claimed to have produced adverse effects, the infused quantities were 2.3–2.9 g [264] and 1.5 g [298] per kg body weight, respectively.

In the xylitol ampoules used in Australia, contaminants were detected. These impurities were analyzed and at least certain imidazole derivatives were found. Their toxicity may have not been sufficient, however, to cause the severe effects observed. The Australian investigators [348] have also strongly claimed that there is no evidence that the xylitol solutions used in Australia would have been someway different from those used elsewhere.

Oxalate crystals in kidneys were also previously observed after infusion of glucose and fructose [350]. Oxalosis therefore also occurs in persons who have not received any xylitol. There is no known biochemical pathway through which oxalate can be formed from xylitol in man. It has occasionally been claimed by certain investigators that glycine could be an intermediate between xylitol and oxalate and that a simultaneous vitamin B_6 deficiency could in part promote such adverse effects. The available literature citations, available by 1976, did not strongly support the role of glycine in the proposed conversation of xylitol to oxalate. On the other hand, a vitamin deficiency, per se, is a severe condition. Numerous dietary ingredients, whether administered orally or intravenously, are not metabolized normally if thiamine and pyridoxine coenzymes are not sufficiently available.

Three fasting subjects of the Turku sugar studies, who consumed 34, 38 and 46 kg xylitol per year, had at the 22-month stage 228, 239 and 465 μmol glycine per 1000 ml serum, respectively. Although these glycine values apparently increased with increasing consumption of xylitol, the values of the three subjects fell within the normal clinical range. Sucrose and fructose consuming subjects exhibited almost similar glycine values (table 7). However, at the 13-month stage of the 2-year trial, the sucrose group (n = 33) displayed a mean value of 183 μmol/1000 ml serum (non-fasting state). The corresponding value for the xylitol group was 348 μmol/1000 ml serum. Consequently, xylitol consumption may have increased serum glycine concentration, but the levels still remained within the physiological range. It would be erroneous to regard such responses as adverse effects. The flexibility of human biochemical system cannot be considered to have been strained to unnecessarily high extent during consumption of xylitol. The amino acid analyses of the Turku sugar studies can be critisized as no analyses were carried out before the commencement of the dietary phase.

Table 7
Concentration of serum glycine and glutamate (in μmol/1000 ml) of subjects who consumed higher quantities of sugars in the Turku dietary studies [302]. Determined at the 22-month stage of the study (fasting state). Sugar consumption is in kg/month.

	Sucrose				Fructose				Xylitol			
Subject No. and sex	1 (f)	16 (m)	86 (m)	Mean[1])	25 (f)	19 (f)	15 (m)	Mean[1])	124 (f)	29 (f)	47 (m)	Mean[1])
Glycine	209	199	222	262	202	294	262	232	228	465	239	250
Glutamate	92	99	128	150	47	90	62	131	138	100	156	202
Sugar consumption[2])	3.3	5.1	3.0	2.2	3.8	3.8	5.2	2.1	2.8	3.8	3.2	1.5

[1]) Mean of the whole test group (n = 30 for sucrose, 33 for fructose and 49 for xylitol).
[2]) Based on diaries. The actual consumptions of sucrose and fructose were somewhat higher [133, 302].

Consequently, it is highly likely that the Australian cases did not result from the use of xylitol in a way identical to that used in Europe where a considerable experience has been gathered about the use of xylitol. Either the way of administration or the quality of xylitol, or both, to patients who already were in a certain critical or abnormal clinical condition, must explain the accidents. Presently exploratory studies are being carried out. Wretlind [351] emphasized in a review that the symptoms described by Thomas et al. in Australia appeared quite individually. He also recalled that the Australian authors made the reservation that the primary disease may be responsible for the reactions.

Related to the Australian cases a statement of Förster should be considered. In a very well-grounded review Förster [352] criticized the study of Schumer [298], because the latter author only published bilirubin concentrations at the end of the infusion without giving initial values. Förster concluded that no liver damage was involved in the study of Schumer. Förster [352] also pointed out that only four out of twenty-two patients in Australia displayed side effects and criticized the use of a misleading title in the paper of Thomas et al. [265]. Förster [352] finally concluded that xylitol is safe and that normal biochemical changes are not synonymous with adverse reactions, and that xylitol may in fact cause fewer changes than fructose or sorbitol.

As a direct consequence of the appearance of the adverse reactions in Australia, studies of xylitol toxicity were carried out by Australian investigators [348]. These infusion studies supported the earlier concept that diuresis, metabolic acidosis and hyperuricemia result from rapid infusion of hypertonic solutions of xylitol. However, glucose, fructose and sorbitol produced similar changes. Xylitol administration was associated with insulin release, decrease in plasma glucose and phosphate, and increase in plasma triglycerides. No hepatic, renal or cerebral damage or calcium oxalate deposition was observed in the dogs studied [348]. Meng [300] in a previous experiment did not observe oxalate crystals in the kidney of dogs given xylitol. The Australian authors stated that the formation of oxalate crystals in their earlier studies was attributed to the possibility that for some reason the metabolism of xylitol did not follow expected pathways. The production of unusual metabolites may lead in some way to oxalate formation. Thomas et al. [265] suggested that glycolaldehyde may be released from the 'active glycolaldehyde' in the transketolase reaction in which this unit is transferred from xylulose 5-phosphate to form sedoheptulose 7-phosphate and glyceraldehyde 3-phosphate. Glycolaldehyde may be subsequently converted to glycolic, glyoxylic and oxalic acid. Thomas et al. [355] have sketched the possible metabolic pathways associated with the formation of oxalate. This outline is shown in figure 7. The question mark in the figure forms a problem. Clearly more studies are required to show the connection between the metabolic steps indicated. It is

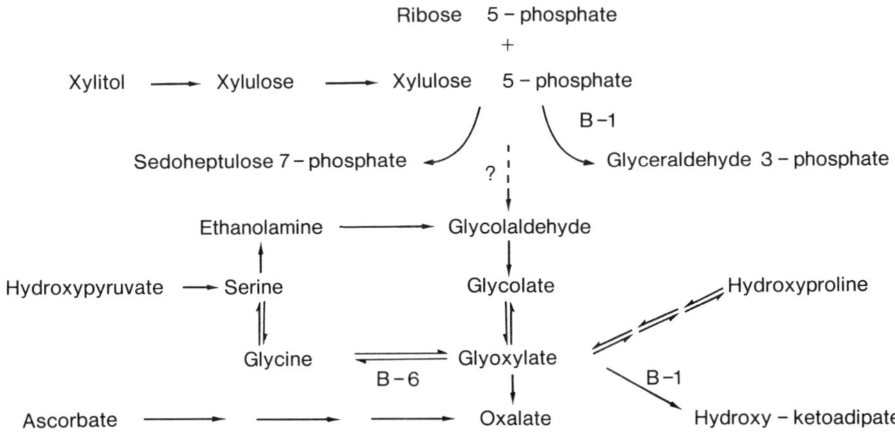

Figure 7
Metabolic pathways associated with the formation of oxalate. The possible connection between xylitol and the oxalate metabolism is indicated with a question mark. The outline was presented by Thomas et al. [355] largely on the basis of rat studies, stimulated by the reported adverse effects of xylitol infusion in Australian patients. B-1 and B-6 stand for vitamins B_1 and B_6, respectively.

well known that oral administration of ethyleneglycol leads to oxalate deposition in kidneys.

Because comprehensive experimental proof for the above reactions has not yet been provided, one is entitled to present other speculative routes of oxalate formation: (1) Heavy intravenous loading of patients with xylitol may temporarily affect the efficacy of transaminations requiring pyridoxal phosphate coenzymes which are responsible for vitamin B_6 activity. Deficiency of this vitamin may lead to impaired transaminations and to an increase of plasma and tissue glutamate and glycine concentrations. It is natural that a rapid infusion of high amounts of glucose, fructose or sorbitol induces similar reactions. Any dietary sugar can finally be an oxalate precursor if transamination reactions are taking place at a reduced rate. (2) Another possibility for the elevated glycine concentrations would be an inhibition in the pathway of purine biosynthesis in the liver. The vulnerable step would be the enzymic conversation of 5-phosphoribosylamine to glycinamide ribotide. For the formation of the latter both glycine and ATP are required (fig. 8). They may accumulate if the catalysis is inhibited. It is likely, however, that in the multiplicity of metabolic reactions several other interconversations are equally likely; a single phenomenon, as the one outlined in figure 8, may not alone come into question. Although the glutamate and glycine values of all xylitol-consuming subjects in the Turku sugar studies were within the normal physiological range, the possible correlation between heavy-consumers and the glycine and glutamate values should now be reconsidered (table 7). The data were not, however, very usable as the amino acid analyses were not performed during the whole 2-year study, including pre-study periods. This matter remains to be elucidated more adequately. It is, however, very unlikely that peroral administration of xylitol would induce abnormal glycine or glutamate values in serum.

A third reason for high serum glycine values is the use in intravenous therapy of sugar solutions supplemented with amino acids. If the solutions contain large amounts of glycine – such brands of infusion solutions have been made – the serum glycine levels will be concomitantly high. The abnormal glycine concentrations may under certain circumstances lead to oxalate formation.

Recently, Brin and Miller [353] also touched upon the Australian adverse effects and expressed the following three possible reasons for these findings: (1) a possible contaminant in the parenteral solutions, (2) dose levels of xylitol, which appeared to be in excess of those found to be safe in other countries, (3) unsuitability of Australian patients for xylitol therapy when compared to European patients treated with xylitol. It can more exactly be said that the Australian experience is not relevant to the safety of xylitol, at least as an oral additive, as the Australian patients suffered from premortal diseases. Thus these patients were unlike the European subjects for whom parenteral xylitol has been used routinely. In Australia the diseases included various conditions which may result in oxalate deposition even without

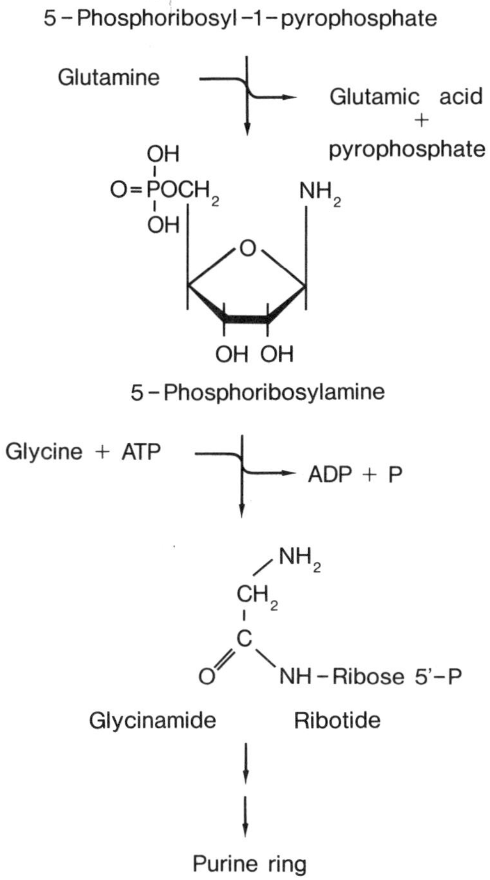

5-Phosphoribosyl-1-pyrophosphate

Glutamine → Glutamic acid + pyrophosphate

OH
O=POCH$_2$ NH$_2$
OH

O

OH OH

5-Phosphoribosylamine

Glycine + ATP → ADP + P

NH$_2$
CH$_2$
C
O NH-Ribose 5'-P

Glycinamide Ribotide

Purine ring

Figure 8
A simplified scheme of the pathway of purine biosynthesis. The scheme involves a step at which accumulation of glycine may occur if the formation of glycinamide ribotide is inhibited.

xylitol, such as ruptured aneurism, intestinal obstruction, crushing syndrome, etc. In Europe, on the contrary, xylitol has been mainly given intravenously in various types of glucose intolerance.

Oxalate has been found to be, according to a literature review, a relatively common histopathological finding in subjects succumbing to various conditions of hemodynamic shock [353]. These conditions have developed without administration of xylitol. Oxalate crystal formation is a very common finding in uremia [354]. Brin and Miller [353] also listed a number of other cases.

Bässler and Schultis [357] have recently stated that it is doubtful whether the oxalate crystal formation, reported by the Australian [265] and German [358] groups, was causally related to xylitol infusion, since the patients involved were gravely ill. Precipitation of oxalate is frequently observed under such conditions [359]. In later studies [360] by Pesch et al. no correlation between oxalate precipitation and the type of carbohydrates used in infusion therapy, was observed. Bässler and Schultis [357] finally concluded that oxalate crystal formation depends on renal condition rather than xylitol infusion.

In a recent symposium on polyalcohols, Thomas [355] presented results of infusion experiments on rats placed on vitamin B_1- or B_6-deficient diets. Rats deficient in vitamin B_6 were also given 5 μCi of U-^{14}C-labelled glucose, fructose, sorbitol or xylitol and the urinary excretions of ^{14}C-labelled oxalate were measured. The excretion of oxalate was significantly increased in vitamin B_6-deficient rats infused with xylitol compared with all other groups. When compared with the control group given xylitol, oxalate excretion was increased 5-fold. Vitamin B_6-deficient rats also excreted increased amounts of glyoxalate and glycine. Only 0.04–0.14% of total radioactivity given was, however, excreted in the form of ^{14}C-labelled oxalate. In the other groups the values were lower than 0.015%. These findings led the investigator to claim that in rats xylitol is an oxalate precursor and that simultaneously an increased excretion of urinary oxalate, glyoxylate and glycine results in vitamin B_6-deficient animals.

The above study has also been critisized with regard to its details, but one of its values is that the report also outlined the possible metabolic pathways associated with the formation of oxalate. Anyway, as indicated previously, heavy loading of patients or test animals with carbohydrates may induce increased formation of oxalate precursors, if an adequate supply of necessary coenzymes for effective transaminations and decarboxylations, requiring vitamin B_6, and for the action of aldehydetransferases, requiring vitamin B_1, are not available. It is clear that the above rat experiment has not indicated that xylitol should be regarded as a special detrimental factor to man. Rapid infusion of almost any nutrient to subjects which already suffer from a very basic nutritional deficiency, requires special attention from a physician. The situation becomes more severe if the patients have premortal diseases. In case of any vitamin deficiency, infusion of carbohydrates should be accompanied by administration of the vitamins involved. The vitamins mentioned may largely be needed in amino acid metabolism of liver. When a vitamin deficiency is involved, the consequences of a prolonged infusion therapy with natural carbohydrates are easy to predict. It was recently shown that a decrease in cerebral transketolase activity (which is dependent on thiamine triphosphate) does not lead to a diminished pentose phosphate cycle activity in thiamine-deficient rats [596].

Several aspects related to safety of xylitol have been listed in table 8. Finally, two conclusive points, also shown in the table, will be examined. The Australian findings about the adverse effects of xylitol administration can be analyzed by simultaneously comparing the separate publications [263, 265, 347, 348]. In certain cases the rate of infusion was indicated to be 0.45 g/(kg×hr) over 24 hours. However, the infusion was performed at three stages during the 24-hr period. The single dosage was thus 8×0.45 g=3.6 g/

(kg × hr), a value which is very high no matter what sugar is administered. It is therefore very likely that the main reason for the adverse effects was too high loading of ill patients with a parenteral carbohydrate, and not a contaminant in xylitol solutions.

The group of Wang and van Eys [268] recently showed with rabbits that no relationship between oxalate accumulation and xylitol toxicity occurred. It is obvious that the reactions of health authorities in various countries which prohibited the import of xylitol, were errors. The import of glucose, sorbitol and sucrose could have been prohibited exactly for the same reasons.

Rat experiments have shown that intake of xylitol is followed by an increase of not only liver glycogen, but also of the concentration of vitamin C, vitamin B_1 (thiamine) and niacin [356]. The latter findings may also partly explain the curative effect of xylitol in conditions of the hepatobiliary system. It is also possible that the administration of xylitol increases the absorption of vitamin A and thereby the concentration of this vitamin in the liver as well [589].

7.32 Other aspects related to the safety of xylitol

Oral administration of high quantities of xylitol does not induce similar changes in blood and urine chemistry as intravenous administration involving same quantities. The fairly slow rate of absorption when compared to glucose and the mere fact that oral consumption of xylitol does not usually involve very high quantities at a time, makes these two ways of administration very different. However, at a suitable infusion rate, xylitol seems to be as safe as glucose or fructose.

Most human or animal experiments involving parenteral or oral administration of xylitol are listed in table 8. A number of these studies were also evaluated by Brin and Miller [353], who concluded that oral administration of xylitol is safe.

A number of reviews have touched upon the metabolism and use of xylitol. Lang [361] discussed the nutritional properties of xylitol and concluded that xylitol is nontoxic and well utilized in intermediary metabolism. Xylitol was also recommended for diabetics. Hötzel [362], Mehnert [363, 364], Förster [352, 365], Bässler [366] and others came to the same conclusion.

Gärtner [367] stated that xylitol is almost universally utilized and is the best carbohydrate for shock, diabetes, ketonemia, liver coma, uremia and postoperative states. In a discussion Birnesser et al. [368] indicated that the adult organism can degrade 10–20 mg xylitol per min and kg. This can be compared to the corresponding metabolism of fructose and corresponds to approximately 80% of the glucose metabolism. Recent very thorough clinicochemical studies of Matzkies et al. [369, 370], for example, strongly support the ideas presented. The overwhelming majority of original and review articles about xylitol emphasize the safety of xylitol administration. Against this background it is interesting to examine the corrective statement of

Froesch [371] that physicians, physiologists and chemists have made an error in considering that glucose could be replaced by fructose, sorbitol or xylitol in parenteral nutrition. The conclusion of Froesch is largely based on the findings that sufficiently high quantities of xylitol and fructose lead to a reduction of liver cytoplasm and to an increased lactate/pyruvate ratio. The considerations presented earlier in this chapter and in table 8 seem to do justice, however, to the opinions presented by Lang, Förster, Mehnert, Matzkies and others. The ability of xylitol to reduce the quantity of free fatty acids in serum when compared to glucose, and to maintain or lower the triglyceride level, seems also to be a generally accepted fact [372].

Förster and Hoffmann [373] are among those numerous investigators who have emphasized the fact that administered glucose is metabolized only by 20–30% by liver, whereas fructose, sorbitol and xylitol are metabolized by more than 80% by this organ. The latter sugars thus influence the metabolic pattern of liver stronger than glucose which is metabolized preferentially by peripheral tissues. In spite of this, there seemed to be only one metabolic effect in which these sugars differ [373]: fructose, sorbitol and xylitol accelerate the formation of urate. Förster and Hoffmann concluded that these three sugars have an advantage over glucose in cases of hepatic carbohydrate intolerance, for example, under stress situations.

Haslbeck [374] also stated that polyols and fructose have advantageous effects because they are metabolized at a rate higher than 80% by the liver with the advantage of an insulin-independent utilization in that organ. The glucose thus finally formed from fructose and polyols is gradually released into the periphery [374]. Glucose raises free fatty acids to 300% from the initial level, whereas fructose, sorbitol and xylitol do not display such effect. According to Haslbeck [374] the rise in serum urate is a specific trait of the metabolism of these three carbohydrates. The parenteral use of fructose and polyols was recommended in diabetes and severe illnesses (stress situation) with an impaired glucose tolerance.

In another comprehensive treatise Förster, Haslbeck and Mehnert [391] emphasized the special importance of fructose, sorbitol and xylitol in parenteral nutrition. This is based on the decreased tolerance of glucose in stress situations (fever, postoperative phase, etc.), as well as in diabetes which generally is not negatively influenced with sucrose substitutes. Additional indications for the use of these sugars are acute pancreatitis and liver failure [391]. These authors have repeatedly stated that the side effects, as increase in serum bilirubin, lactate and urate, and decrease in liver adenine nucleotides, have no practical significance in reasonable therapy. Combined application of sugars and sugar alcohols was considered to be a valuable future method in parenteral nutrition [391]. The value of sugar combinations was further indicated by Geser [392] who also recalled the fact that in stress situations glucose utilization is diminished. Intravenous glucose administration at a high infusion rate can lead to hyperosmolar hyperglycemic coma. Fructose, sorbitol and xylitol do not essentially modify glucose homeostasis [392].

The findings of Woods and Krebs [375, 376] that xylitol loading may cause a depletion of ATP, adenine nucleotides and inorganic phosphorus, were obtained in liver perfusion experiments. These results were largely similar to those of Brinkrolf and Bässler [377], obtained in rat infusion tests. The former authors regarded their rat liver perfusion tests as adequate to claim that infusion of xylitol in man is not without risk. This is apparently an overestimation of in vitro animal studies and also contradicts the earlier in vivo results [377] on rat. However, fructose and sorbitol also seem to decrease ATP, total adenine nucleotides and inorganic phosphorus in the liver, the effects being dose-dependent [378, 579]. This was considered to be due to the rapid phosphorylation of fructose and sorbitol. ATP and inorganic phosphorus are inhibitors of AMP breakdown. When the levels of these compounds are reduced, AMP is said to be degraded to urate, the final product of AMP metabolism [378]. This has been regarded as the cause of hyperuricemia. Heuckenkamp [586] was also inclined to believe that infused xylitol may induce the formation of new nucleotides via de novo purine biosynthesis. In the de novo biosynthesis of purines the rate-limiting substrate is 5-phosphoribosylpyrophosphate (PRPP). There is a report [587] which claimed that infusion of xylitol at a rate of 0.4 g/(kg$\times$hr) increases the generation of PRPP in erythrocytes. Heuckenkamp concluded that due to the high urinary losses of xylitol at an infusion rate of 0.5 g/(kg$\times$hr) and the remarkable hyperuricemia observed during xylitol infusions ‘physicians should be advised against the use of this pentitol in metabolically healthy subjects’.

In a liver perfusion test, Jacob et al. [288] observed only slightly diminished adenine nucleotide levels and the ATP/ADP-ratio was not appreciably changed. In another type of in vitro studies it was shown that addition of xylitol to stored blood resulted in regeneration of ATP and other high-energy phosphates [279, 311]. In a rat testicular homogenate system both glucose and xylitol increased the levels of ATP [379]. According to Brand and Quadflieg [567] the decrease of the content of ATP and 2,3-diphosphoglycerate in human erythrocytes is not dependent on the kind of sugar used, but related to the amount of carbon atoms taken up. The experimental conditions in the above studies were not identical (for example liver cells versus erythrocytes), making a direct comparison between the studies difficult. The response of adenine nucleotide levels as affected by xylitol seems to be, however, inadequately investigated. It is also likely that too strong claims were presented on the basis of the rat liver perfusion tests [375, 376]. The biochemical findings about the decreased adenine nucleotide values in these experiments were valid, but the claims that these changes would be the reason for adverse xylitol effects in man can not yet be regarded as conclusive. An addition to this discussion was provided by Fekl [581] who stated that under steady-state conditions during infusion one does not find an ATP-decrease (shown initially by Bässler). Fekl has observed the ATP-fall only directly after giving a sudden load of substrate which has to be phosphorylated. The fall was not found later. If an ATP-fall occurs, it depends on the substrate concentration and the activity of the required kinase [582]. It should also be noticed that a loading of the kidney with glucose leads to a 40–50% reduction of ATP from the original value, although the hexokinase is more active in the kidney than in the liver [582]. The ATP-fall in the liver can be avoided by considering the substrate concentration. At a rate of 0.25 g/(kg × hr) no decrease in the concentration of ATP does occur in continuous infusion [582].

Horecker [395] showed in a review that pentoses and pentitols have a decisive and important role in the biochemical evolution. Their significance should not be overestimated either. A critical survey of available literature reveals, however, that xylitol has a considerable value in human nutrition. A wise application of xylitol may have useful health effects and such should be exploited.

7.4 Nutritional aspects

Xylitol has been used on a large scale for parenteral feeding for several years. In the past years the xylitol pharmaceutical market in Japan has been increasing strongly. The bulk of this xylitol has been used in parenteral nutrition. In 1972 more than 364 tons of pharmaceutical grade xylitol was sold in Japan alone [14], and considerable quantities in West Germany as well. More than twenty million infusion bottles were used by the middle of the 1970's. Xylitol has been successfully used in parenteral nutrition mostly in cases in which glucose utilization is impaired. For intravenous infusion, 100–400 g of xylitol has been given daily at a rate of 0.125–0.5 g/(kg × hr) [521, 535]. According to the present understanding (table 8), xylitol causes only minimal insulin secretion in comparison with other dietary carbohydrates [380]. It should be noticed, however, that xylitol, in high doses (more than 0.8 g/kg) enhances the release of insulin [578], and that glucose, fructose and mannose also stimulate the insulin release. The membrane transport of xylitol is independent of the presence of insulin. These two properties have made xylitol suitable in many applications of parenteral nutrition. Other important properties include the fact that the absorption of xylitol is usually much lower than that of glucose, but the utilization of xylitol is very high [289]. Xylitol infusion may lead to slightly lowered blood glucose levels (table 8). These values increase to control levels after the infusion is terminated [381].

Detailed studies [300] on beagle dogs indicated that it was difficult to detect any difference between the use of glucose and xylitol as a carbohydrate source in parenteral nutrition, when the daily infusion of xylitol was limited to 14–20 days at a daily dosage of not more than 10 g/(kg × day). Meng [300] emphasized that the rate of infusion of xylitol, as well as the length of infusion period, are of paramount importance to achieve maximal utilization and to avoid adverse effects. Too concentrated xylitol solutions (20–50%) were not well tolerated in rabbits and dogs [300, 382]. Each animal species possesses its own specific toleration range and human subjects differ in this sense from each other to a certain extent as well. This concerns most nutrients administered orally or intravenously. Consequently, the effects resulting from the intravenous infusion of xylitol, as increase of serum lactate resulting in lactate acidosis [298, 347, 383], and increase of pCO_2 [298, 347], should also be judged in light of the above considerations. Meng [300] did not observe lactic acidosis and any significant change in blood pH in dogs given large dosages of xylitol. Other sugars, as glucose, may exert similar effects as xylitol [383].

Related but likewise explainable discrepancies have existed as regards elevation of plasma insulin following intravenous administration of xylitol. Plasma insulin increased in patients [384] and dogs [385] given xylitol intravenously in a short-term experiment, but not in dogs which were infused continuously during a 24-hour period [300]. Meng [300] concluded that a mixture of two or even three sugars should be used in total parenteral nutrition. Glucose should be a constant component. Fructose, sorbitol, xylitol or fructose-xylitol mixtures may be given in combination with glucose. It was further established that long-term administration of xylitol either singly or in combination with amino acids, vitamins and minerals in total parenteral nutrition at dosages of 8–16.75 g/(kg×day) and with an infusion rate of 0.35–0.7 g/(kg×hr) is apparently tolerated and utilized for energy in dogs. In Germany mixtures of various sugars have been successfully tested in parenteral nutrition in man. Bickel [557] further stated that if glucose, fructose and xylitol are infused in the immediate postoperative period at rates necessary for parenteral nutrition, all these substrates produce specific undesired effects. These effects can be reduced significantly when the three substances are infused in the form of a mixture. The composition and the rate of such a mixed solution are important [557]. The advantage of the carbohydrate mixture is that the three substrates are metabolized via primarily independent metabolic pathways; mutual inhibitions of turnover do not occur [557].

Bässler [577] pointed out that there is no formula for the calculation of suitable and safe infusion rates for sugars. It is natural that the rates should be determined empirically for each carbohydrate. Bässler [577] gave recommendations (largely based on the data of Bickel and Berg) for safe infusion rates, but emphasized that the safety depends on the number of parameters considered. Accordingly, for xylitol the upper limit of safe infusion rate was considered to be 0.25 g/(kg×hr), depending on the metabolic situation. The mean of maximum metabolic capacity for xylitol was calculated to 0.5 g/(kg×hr) in postoperative patients and to 0.37 g/(kg×hr) in healthy subjects. The upper limit for a mixture of fructose, glucose and xylitol (2:1:1) was 0.5 g/(kg×hr). Bässler [577] also stated that infusion of mere carbohydrates may lead to metabolic disturbances which can, however, be avoided to a certain extent by simultaneous application of electrolytes and amino acids. Evaluation of the infusion experiments with carbohydrates entitles to the question why the importance of electrolytes, amino acids and vitamins was not realized earlier.

The 2-year feeding study in Turku showed that xylitol has several nutritional advantages which suggest that it can be applied as a non-cariogenic, cariostatic and therapeutic sweetener, both for diabetic and other subjects.

Other uses may include application as sugar substitute for obese persons. There is a considerable amount of information suggesting that xylitol may contribute less to the formation of fatty tissue than the amount of sucrose equivalent to its sweetness [384, 386–390], a matter which was one of the main findings of the first major symposium on xylitol [290]. The weight of the subjects of the 2-year Turku feeding studies remained practically constant regardless whether the whole xylitol group (n = circa 50) or females and males, or subjects (n = 17) consuming the highest quantities of xylitol were separately inspected. The group was apparently too heterogenous for accurate weight control studies. Free of charge distribution of foodstuffs slightly increased the total carbohydrate consumption of several xylitol subjects. Six male and female subjects, aged 23 to 48 years, displayed approximately 5% weight reduction already within a few months after the onset of the dietary regimen. These subjects used prior to the study, during its course, and after its termination, approximately the same daily doses of foodstuffs. In all cases the weight returned to the pre-study level when sucrose-consumption was again commenced. These six subjects did not have a major effect on the mean of the whole group.

The Turku sugar studies were characterized by the consumption of a very varied xylitol diet [133]. The versatile experience gathered from these experiments indicates that xylitol is very well suited for various food-manufacturing purposes. Its behaviour in food processing made it suitable and, in numerous cases, an ideal sweetener in bakery products, syrups, squashes, canned food, jams, jellies, marmalades, confectionaries and pharmaceutical products. Replacement of sucrose with xylitol led, according to interviews carried out in Turku during 1972–1975, to very appealing chocolate products. chewing gums and many other products. Consequently, the long-term dietary studies in Turku are strongly inconsistent with the single coffee-break test mentioned by Froesch [588] who reported that 'xylitol chocolate was horrid'.

7.5 Ignored facts

The literature referred to in chapter 7 indicates that, as regards the use of xylitol, sorbitol, fructose and other carbohydrates in nutrition, one is faced by the following familiar situations:

- A chemical compound which has been shown to be a very useful tool in the hands of one physician, is dangerous and toxic in those of another physician. The main reason for the discrepancies is not in the differences of the empirical data; they are essentially consistent. The question is how the data should be interpreted.
- Results of in vitro animal experiments have been overestimated to a certain extent. Such results should not as such be applied to man.
- Several investigators have ignored the fact that differences between animal species are understandable and expected. The use of dog, for example, as an experimental model is valuable, if it is realized that the

results achieved in this model cannot be directly applied to hundreds of other mammalian species. Man has partly by accident acquired his domestic and experimental animals. Dog, rat and rabbit only represent selective cases. Even primate species differ considerably from each other as regards their endocrinology and exocrinology. Certain monkeys (species of the genus *Macaca,* for example) are not at all good models for humans as far as the salivary gland physiology is concerned.

- Some pathologists have ignored the fact that in long-term toxicological studies pathological changes regularly develop in old rats and mice on various onesided diets containing high amounts of normal dietary ingredients such as fats or various carbohydrates, including sucrose, glucose or sorbitol. Ingestion of high quantities of such common chemicals or foods as table salt, aspirin, citric acid, ethanol, coffee, tea, chocolate, mustard, etc. most likely induces pathological changes in most animal species, if the high consumption continues for the whole life span.
- Parenteral and enteral administrations are basically very different ways of giving any nutrient, but this fact has also been ignored to a certain extent. The only physiological route is through the whole alimentary tract. Furthermore, although it has been repeatedly emphasized that the use of vitamins and trace elements in parenteral nutrition should always be considered in connection with the administration of carbohydrates [393], these facts are sometimes ignored.
- Any ingredient in the human diet induces adverse effects if its concentration is brought to a sufficiently high level. All effects are dose-dependent.
- There is not a diet which would not change the host. Diet has chemically affected man throughout the evolution and this situation will continue. This is natural and inevitable. One should be able to differentiate the understandable or physiological changes from the pathological ones.
- It has also been partly ignored that pronounced individual differences exist between human subjects as a result of age, nutrition, underlying diseases and other factors. Lactose and glucose, for example, are not at all suitable dietary carbohydrates to millions of people.
- The fact that fructose, sorbitol and xylitol appear to be, according to most studies, suitable in parenteral and enteral nutrition, does not lessen the firmly established value of glucose in the same purpose. Unnecessarily created competition does not alter physiological qualities of dietary sugars and they can thus be used depending on the cases involved. It would be wrong to deny the benefits of glucose either, for example, such as mentioned by Heuckenkamp and Zöllner [394].
- The consumption values of xylitol in the Turku sugar studies should not be misunderstood and misused. The highest daily dosages of xylitol, viz. 200–400 g, are not realistic with regard to most future applications of xylitol in man. Such amounts would come into question only in restrict-

ed and controlled cases. As these high intakes did not result in any pathological symptoms, it is likely that much lower intakes, viz. 10–70 g daily, will not produce such symptoms either.

8. Conclusion

Xylitol is an intermediate of carbohydrate metabolism in man, animals, many plants and microorganisms. In animals it occurs in the glucuronate-xylulose cycle. Most fruits, berries and other plant material contain xylitol.

Xylitol is easily converted into glycogen, glucose and its metabolites in the intermediary metabolism of man. This, and the fact that enteral xylitol is absorbed more slowly than glucose, lead to a low steady-state xylitol concentration in human blood. The normal xylitol concentration of blood is 0.03–0.06 mg per 100 ml. Xylitol is excreted in normal human urine at a rate of 1.0–2.5 μmol/hr. In diabetics the excretion may vary from 1.0 to 4.0 μmol/hr.

Xylitol is absorbed passively in man approximately at the same rate as sorbitol and the absorption is practically complete. Enteral xylitol may cause in some unadapted subjects transient osmotic diarrhea in dosages of 0.5 g/(kg×day). Most unadapted adult subjects can consume 30–60 g xylitol per day without side effects. After adaptation, human subjects have taken in 200–400 g xylitol daily without side effects which would differ from those caused by equivalent amounts of fructose or sucrose. In children, aged 3–17 years, no side effects have been observed after 4.5-year chronic intake of xylitol comprising daily doses from 3 to 50 g. The urinary excretion does not then exceed 1%. Xylitol is better tolerated than sorbitol.

In man, approximately 80% of administered xylitol is metabolized in liver, 20% being metabolized in extrahepatic tissues, viz. kidney, erythrocytes, heart muscle, adipose tissue, adrenal cortex and others. These values may differ from those obtained with certain experimental animals. The rate-limiting step in the degradation of xylitol in human liver seems to be the NAD-xylitol dehydrogenase-catalyzed reaction. Endogenous and exogenous xylitol is metabolized in the glucuronate-xylulose pathway. The final products are the same as after intake of glucose.

Endogenous xylitol is produced from D-xylulose or L-xylulose in mitochondria. Exogenous xylitol is dehydrogenated in the cytoplasm. In the former case the xylitol-specific dehydrogenases require either NAD or NADP. In the latter case a non-specific polyol dehydrogenase (iditol dehydrogenase; EC 1.1.1.14) is involved. When human subjects become adapted to use high quantities of xylitol, there is an increase in the amount of liver cytoplasm polyol dehydrogenase.

Xylitol can be considered to have three levels of entry into the glycolytic pathway. Sorbitol has one. In contrast to xylitol, the dehydrogenation of sorbitol by polyol dehydrogenase is not the rate-limiting stage. Among the parenterally used carbohydrates xylitol seems to be the only carbohydrate which can completely replace glucose in the glucose-6-phosphate dehydrogenase-catalyzed reaction to produce pentoses, because it can enter the pentose phosphate pathway independently of this enzyme.

Xylitol, as well as fructose and sorbitol, is converted primarily in liver into glucose and its metabolites. Thus xylitol is finally liberated into circulation as glucose and either stored as glycogen, oxidized to water and Co_2, or utilized in the biosynthesis of body constituents, using normal routes of metabolism.

Xylitol has been infused in healthy human subjects at a rate of 0.125–0.250 g/(kg × hr) without any side effects. In many cases no side effects have been found at a rate of 0.5 g/(kg × hr). A critical review of available literature indicates that the changes of certain blood values within physiological range have erroneously been regarded as adverse effects. As a result of the slower absorption of xylitol, the metabolic capacity is never exceeded in enteral administration which takes place on rational grounds. Patients in shock seem to have a greater metabolic capacity to utilize xylitol than healthy subjects. Surgical patients seem to have a lower urine excretion rate of xylitol than healthy subjects. In infusions with infants, a metabolic capacity of 0.5 g/(kg × hr) has been determined. In parenteral nutrition xylitol seems to be as valuable as glucose. Recent development indicates that mixtures of xylitol with certain other carbohydrates (fructose, sorbitol, glucose, etc.) will play an important role in infusion therapy.

Xylitol is the only natural carbohydrate which seems to meet almost all requirements set for a sucrose substitute in human diet. Xylitol is at least non-cariogenic in man, but it most likely has an active property to fight dental caries as well. From the dental and physiological point of view, xylitol seems to be safer and more advantageous than sorbitol. The advantages of xylitol are based on a higher sweetness/calorie ratio, non-fermentability by cariogenic bacteria, stronger ability to reduce plaque growth, higher tolerability in alimentary nutrition and more pleasant organoleptic properties, when compared to sorbitol. When compared to xylitol, various hydrogenated hydrolysis products of starch, often used as sweeteners, have the disadvantage of having complex and varied chemical composition.

Xylitol, when being brought into the metabolism of man independently of insulin, has been considered a good sucrose substitute in diabetes.

Xylitol can be regarded as a non-toxic natural carbohydrate. At high loading with glucose, fructose, sorbitol or xylitol similar metabolic changes are observed. The only main differences between these sugars seem to lie in the dosages which are required to elicit the individual changes. Such reactions are not termed adverse effects, but normal responses of the biochemical system of man.

Xylitol is very well suited for various food-manufacturing processes. The negative heat of solution and higher sweetness compared to other sugar alcohols, make xylitol valuable in food and confectionary industry.

No objective scientist with a sense of responsibility would recommend the application of a chemical compound if there is a risk of harmful or adverse health effects. Virtually all scientists are anxious to warn as they would then be on the safe side if anything negative happens. This is usually acceptable. The hackneyed and routine warnings may, however, sometimes quench the wise, careful and controlled use of a promising natural substance. Hence it requires courage to stay behind findings and claims of a positive nature.

The list below presents an outline of certain more or less current, and suggested future applications of xylitol. While a number of them still require more thorough investigations, many of them clearly have a current utility value.

(1) Oral administration:

- As a sweetener and source of energy for subjects with rampant dental caries, or caries-prone subjects, particularly in low-fluoride areas. As a general non-cariogenic agent. Reduction of juvenile, adult and senile caries.
- As a sweetener and source of energy for patients with impaired salivary gland function or with reduced salivary flow rate (as a result of genetic reasons, evisceration, radiation therapy, Sjögren's syndrome, etc.).
- Enhancement of the peroxidatic capability of secretions and cells containing peroxidase, as in the gut wall (for treatment of colitis?) and salivary and mammary glands.
- As a sweetener in diets for patients with oral candidiasis (the ability of *Candida* cells to use xylitol should, however, be separately checked in each case).
- As a sweetener for subjects wearing orthodontic appliances which accumulate dental plaque.
- As a possible gingivitis-depressing agent. As an agent which in long-term use would partially reduce the development of alveolar bone loss and other consequential conditions. As an agent which indirectly alleviates inflammation in the oral cavity.
- As an agent increasing the hydrolysis of food in the oral cavity during mastication and certain period after swallowing (through the increased activity of proteinases, amylase and certain other salivary hydrolases).
- In general, as an agent increasing the efficacy of natural salivary and other exocrine defence mechanisms. In addition to the use of xylitol as a sweetener in diet it could be administered orally or topically in prophylactic, non-cariogenic or remineralizing gums, lozenges, mouth washes, tooth sprays and pastes, and used as a sweetener in medicines like cough mixtures, tonics, antibiotic syrups aimed for children during treatment

of chronic diseases, and in other pharmaceutical products. Thick xylitol syrups could be used in special applicators for local treatment of the gingival tissue or plaque-harbouring areas.

- As a sweetener and source of energy for diabetic, uremic and certain other subjects with pancreatic, hepatic and renal diseases.
- In therapy of syndromes involving glucose 6-phosphate dehydrogenase deficiency of red blood cells. In addition to oral therapy, intravenous administration may come into question (in neonatal jaundice, for example).
- As a sweetener and source of energy in certain carbohydrate intolerances.
- Application in polyol mixtures in medicated food to treat acetonic disease of cattle.
- Polyol mixtures containing xylitol can also be used as fodder and as a vehicle or preservative of vitamins, etc. in animal feeding.
- Due to the minimal fermentation of xylitol by most yeasts, molds and bacteria, viscous xylitol syrups may be suitable for thickening, sweetening and enriching of many foodstuffs in which microbial growth must be eliminated or reduced. The use of xylitol as a physiological sweetener may considerably increase the shelf life of such products.
- The following potential applications should also be considered: treatment of fatty liver, stimulation of impared detoxification capacity of the liver, enhancement of the absorption of iron and certain other elements and vitamins. The theoretical background of xylitol research entitles to these considerations.

(2) Infusion therapy (either with solutions of xylitol alone or with mixtures containing glucose and fructose). For example:

- Post-traumatic (heavy burns) and postoperative catabolism when the impairment of glucose utilization cannot be reversed by insulin.
- Impaired utilization of glucose in subjects with certain renal diseases. Treatment of uremic patients.
- Resuscitation from diabetic coma. Treatment of diabetes mellitus with intravenous doses of xylitol. Treatment of patients with certain other pancreatic diseases and liver failure. Rat studies suggest that xylitol should be tested in the treatment of conditions of the hepatobiliary system in general.
- Treatment of patients after myocardial infarction and related conditions characterized by glucose intolerance and insulin antagonism.
- Certain disorders in lipid metabolism. Reduction of the concentration of free fatty acids in blood.
- Corrections of conditions characterized by a deficiency of NADPH. For example, if a disturbance occurs in the 'normal' direct oxidative pathway of glucose 6-phosphate to 6-phosphogluconate, as a result of deficien-

cy of NADPH, additional NADPH could be obtained from the pentose phosphate pathway of xylitol metabolism.

- Enhancement of impaired peroxidatic capacity in certain tissues or exocrine secretions of the body.
- Regeneration of sympathetic nerve tissues, which is mediated by a peroxidase. Regeneration is possible if the pentose phosphate pathway is working. Stimulation of this pathway through xylitol administration may be possible.
- In the storage of red blood cells and blood. Red blood cells stored in xylitol are functionally ready at blood transfers.

New important pieces of information about xylitol are obtained monthly and the research of xylitol is at present in an exponential phase. It is likely that new potential applications of xylitol will soon be found in the field of liver, pancreas, bile and gastrointestinal therapy. Certain properties of xylitol may automatically lead to such discoveries. The stimulation of the glucuronate-xylulose cycle by xylitol in the liver may enhance the function of the tissue as a whole in certain liver diseases. The overall detoxification rate of toxic compounds or drugs in the liver may also increase. The ability of xylitol to form chelates with certain metals may affect the absorption of these elements in a positive way. The fact that xylitol is not very readily attacked by gastrointestinal microorganisms should be effectively utilized in practical applications in both humans and domestic animals.

9. Addendum

During the preparation of the present survey, a number of xylitol studies have been published. Most of them will be discussed in the subsequent three chapters.

9.1 Physiological and dental aspects

Mühlemann et al. [623] showed that the cells of *Streptococcus mutans* (OMZ 176) and *Actinomyces viscosus* (OMZ 105) did not ferment xylitol, but xylitol did not affect the growth of the cells. It was further found that xylitol did not interfere with the utilization of sucrose by the above species. Xylitol added to starch or sucrose diets did not affect the formation of bacterial agglomerates on rat molars. In a previous study [110] it was observed that xylitol had no strong effect on the uptake of glucose by the cells of *Str. mutans* (Ingbritt). However, 3% xylitol inhibited by 40–50% the activity of dextran-hydrolyzing extracellular enzyme(s) in the growth medium [110].

In the above study [623] it was further found that rinsing with 10% xylitol solutions did not interfere with early plaque formation in young adults. There were no significant differences in early plaque formed within 72 hours among sucrose and xylitol rinsing treatments (six times daily for 1 min with 10 ml of the test solutions). In previous studies 10% xylitol rinses (five times daily, 100 ml rinse for 1 min), combined with the use of small amounts of xylitol in coffee or tea, produced significantly lower plaque scores than corresponding sucrose solutions [127]. Mühlemann et al. [623] showed, however, 10% xylitol rinses and xylitol chewing gum to be non-acidogenic, a matter which has now been proved by numerous studies. However, xylitol was not found to prevent 'rapid sucrose glycolysis'. In a simultaneously performed rat caries experiment xylitol was non-cariogenic [623]. The rats drank less when xylitol was added to the drinking water. This has not been found in the Turku Institute, when the concentration of xylitol was kept on levels of 2 to 6%. The animals drank xylitol-sweetened water as much as tap water. However, the same animals drank up to 2-fold volumes of water sweetened with 2–6% glucose (Mäkinen, unpublished results).

Three xylitol reports were read at the 55th Session of the International Association of Dental Research in Copenhagen in 1977. Single doses of xylitol- or sucrose-containing chewing gums did not differ significantly with respect to the activity of certain whole saliva enzymes immediately following stimulation (six samples collected within 32 min) [607]. The enzymes included lactoperoxidase, α-amylase and carbonic anhydrase. It was suggested that a part of the salivary HCO_3^- ions may be translocated from blood and/or the extracellular compartment through the oral mucosa. The activity of carbonic anhydrase was analyzed in this study with a microdiffusion method in which HCO_3^- ions served as substrate. Although the method may be applicable in the case of more simple biological samples, it is perhaps not as suitable or successful when used for whole saliva analysis. In later studies (Mäkinen et al., to be published) refined methods of analysis were employed in studies on human parotid saliva. In this case a saturated CO_2 solution is used as substrate and the enzyme activity is determined within a matter of seconds as is necessary for observing a predetermined pH drop in the reaction mixture. This drop is based on the fact that in the enzyme-catalyzed hydration of CO_2, according to the reaction, $CO_2 + HOH \rightleftharpoons H_2CO_3$, the addition of Veronal® buffer disturbes the equilibrium of the H_2CO_3 formed by the introduction of OH^- ions. The concomitant pH change can be measured. Using this method, stimulation of parotid saliva with fruit pastils sweetened with xylitol was observed to be associated with higher initial activity of carbonic anhydrase compared with sucrose-sweetened pastils. Consequently, there was an apparent correlation between the increased salivary HCO_3^- concentration obtained immediately following stimulation with xylitol chewing gum in a previous study [568] and the later carbonic anhydrase findings. It was only possible to observe this correlation with the first few milliliters of parotid saliva collected after stimulation. However, the individual variations of the salivary carbonic anhydrase activities are so high that this study needs verification.

The previously mentioned carbonic anhydrase study [607] also showed xylitol chewing gum to result in higher whole saliva pH values (mean 7.75) than sucrose gum (mean 7.45). These values were measured 10 minutes following stimulation.

Poole and co-workers showed the activity levels of parotid saliva lactoperoxidase to be higher following stimulation with xylitol-containing fruit pastils (also containing sorbitol) than was the case with sucrose pastils [606]. The authors were inclined to believe that lactoperoxidase is acutely induced by xylitol and that this is one of the mechanisms for xylitol-induced anti-caries activity, much in line with suggestions made in other investigations [223].

The effect of xylitol feeding on the incidence of fissure caries was investigated in a rat experiment [624]. Xylitol feeding was associated with lower caries scores than feeding starch alone or a starch-sucrose mixture.

An International Symposium on Xylitol was held in London in May 1977. The main emphasis was put on dental aspects, but other characteristics of xylitol were also discussed. Most of the information presented was not original: the occurrence, manufacture and properties of xylitol were reviewed

[625]. The first report of this Symposium stated, among other things, that 'xylitol cannot realistically be considered as a competitive threat to common sugar. First of all, the dental trials at the University of Turku demonstrated convincingly that a complete substitution of xylitol for sucrose in the diet is not necessary to bring dental caries under control' [625]. The article further listed some physico-chemical properties of xylitol. The value of xylitol as an ingredient in confectionery was also discussed [626]. This article dealt with various applications of xylitol in the household and food industry, with current information concerning the legal status of xylitol in various countries. Related aspects had been previously discussed by Kracher [15]. Kracher further touched upon these points in the above-mentioned symposium [627].

Bässler [628] discussed the biochemistry of xylitol emphasizing the discovery of the glucuronic acid-xylulose cycle by stating that the cycle presumably has the function of recycling glucuronic acid, not used for synthetic reactions or biotransformations, into glucose metabolism. In the subsequent report of Förster the tolerance in the humans for xylitol was examined, with special emphasis on the resorption of carbohydrates [629]. The important characteristic of xylitol and sorbitol, as assessed on the basis of the current knowledge, was repeated: these polyols have not been shown to be absorbed by any special transport mechanism. The resorption of both of these polyols would appear to take place exclusively by means of free diffusion, or by as yet undiscovered low-affinity transport systems.

The adaptation of humans to increasing amounts of xylitol was discussed by Förster [629]. This phenomenon was actually first reported by Bässler et al. [429]: in animal-feeding experiments a sort of 'hardening' to xylitol occurred. The animals were able to tolerate significantly greater amounts of xylitol when the dose rate was gradually increased, than when a larger quantitity of xylitol was administered from the outset. It was suggested that this adaptation is associated with an enzyme induction. It is postulated that the enzyme which catalyzes the initial oxidation of xylitol (xylitol dehydrogenase or sorbitol dehydrogenase), is stimulated under the influence of chronic administration of xylitol. This would result in a more rapid conversion of the xylitol reaching the blood from the intestines resulting in turn in a higher concentration gradient of xylitol between lumen and plasma (compared to the non-adapted organism) and hence a higher rate of absorption. According to Förster [629], however, the mechanism of this theory, has not been verified, and the true circumstances still await elucidation.

Förster cited the fact that even with high oral doses of polyols, their concentration in the peripheral blood hardly ever rises above 5 mg/100 ml, and he viewed it as unlikely that changes in such small concentrations could result in concentration gradient differences sufficient to explain the marked differences in tolerance between adapted and non-adapted organisms [629]. The above-mentioned adaptation to gradually increasing amounts of xylitol was, however, also established in the Turku feeding studies on humans [133].

Concerning the tolerability of xylitol in man, Förster [629] reiterated the quantities of xylitol which do not produce any side-effects: A healthy person can tolerate 10–30 g of xylitol in a single dose without side-effects. In children, no negative effects occur with a single dose of 10 g of xylitol. These are, however, very conservative estimates. It is known, for example, that subjects studied in Turku comprising both adults and children, have consumed single xylitol doses which were higher than those mentioned above without any side-effects [133, 614]. Förster [629] also stated that adult subjects may be able to tolerate more than 200 g of xylitol per day after habituation on graduated dosages. As cases of such high consumption are admittedly rare, it can nonetheless be safely stated that daily consumption of 50–70 g of xylitol can be tolerated by virtually all adult subjects without problems.

A further report of the London Xylitol Symposium dealt with biochemical findings in exocrine secretions in relation to peroral administration of xylitol [630]. The secretions studied included saliva (parotid, submandibular and mixed saliva), milk and lacrimal fluid. Xylitol administration slightly increased the activity levels of certain salivary enzymes and the concentration of protein in saliva, but such changes should be understood as normal and physiological responses to changes of the composition of diet. The enzymes which were studied in exocrine secretions following peroral administration of xylitol included lactoperoxidase, α-amylase and carbonic anhydrase. The article suggested that research should be extended to comprise the whole exocrine system, as xylitol and certain other sugar alcohols may exert selective effects to a greater extent than previously assumed.

Microbiology of the oral cavity and dental caries was discussed by Gehring [631] who concluded that 'xylitol without doubt is the best sugar substitute'. Gehring also reported on studies of the adaptation of a *Streptococcus* species (RX1) isolated from the oral cavity of the rat. Acid production from xylitol did not increase after 2 years of exposure to xylitol and possibility for adaptation, suggesting that fermentation of xylitol by xylitol-degrading cells is genetically fixed. Later studies have shown no increase in the acid production from xylitol in dental plaque after 3.2–4.5 years' chronic consumption of xylitol by humans [614].

Gehring [631] further discussed the fermentation of xylitol by plaque flora obtained from the Turku sugar studies. Although the acidity produced in the presence of xylitol by the plaque flora were hardly ever below pH 6.5, there were a few isolated cases in which a pH value of 5.5 was recorded. These values occurred sporadically, however, in all three experimental groups (sucrose, fructose and xylitol), and at the end of the two-year study by no means more often in the xylitol group. Gehring isolated a few xylitol-fermenting bacterial strains from these rare cultures of plaque flora cultivated in the presence of xylitol. The strains were identified as streptococci, but it was not possible to associate them to one of the known serological groups A–S of streptococci [60]. These bacterial strains and similar streptococci that were isolated from rats' oral samples, showed the characteristics of the serological streptococci group Q which may be the only streptococcal serogroup for which a positive xylitol degradation has been described. The newest edition of Bergey's Manual [632] lists *Streptococcus avium* as such an organism. It mainly occurs in the feces of chickens and occasionally in humans, dogs and pigs. According to Gehring, germs of the family Enterobacteriaceae from the oral cavity of the rats and not yet identified gram-negative bacilli from the oral cavity of the hamster, show xylitol degradation [60, 631]. According to Bergey's Manual xylitol can also be fermented by a single *Lactobacillus* species, *Lactobacillus salivarius* ssp. *salivarius.* However, this statement has not so far been confirmed. *Staphylococcus saprophyticus* has also been mentioned among xylitol-metabolizing microorganisms [633]. Gehring finally states that the ability of microorganisms to metabolize xylitol compared with the microbial utilization of hexoses seems to be rare, particularly with regard to the flora of the oral cavity of humans [631].

The 24th ORCA Congress (Organisation for European Caries Research) held in Megève, France, in 1977 included a few reports on xylitol. In a study which compared the immediate effects of several sweeteners, it was shown that 10% xylitol rinses increased salivary pH more than sucrose, glucose or even water [634]. This rinsing did not cause any significant differences in the activity levels of whole saliva lactoperoxidase, nor in the concentration of inorganic phosphate. Mechanisms which were responsible for buffering and pH-rise, were affected within two minutes following stimulation with xylitol. It is unlikely that oral microorganisms were responsible for this effect, but rather the major or minor salivary glands, or both.

Havenaar et al. [635] compared the ability of several plaque bacteria to metabolize different carbohydrates. Altogether approximately 200 strains from over ten species were tested. None of the strains could ferment xylitol. No adaptation to xylitol was detected either. Several bacteria could ferment lactitol. In further comparisons, Gehring and Karle [636] found sucrose more cariogenic than sorbose which was more cariogenic than xylitol. These authors suggested that 'L-sorbose possibly is a low-cariogenic sugar substitute, but that among all sugar substitutes known to date, xylitol still seems to be the most advantageous in caries prophylaxis'. Büttner and Moll [637] showed that even partial substitution of xylitol reduced the cariogenicity of sucrose in rats.

9.2 Aspects in parenteral nutrition

The International Society of Parenteral Nutrition held its second meeting in Kyoto, Japan, in 1975. Several reports were presented which dealt with the use of xylitol in parenteral nutrition. Lee and Wretlind [638] discussed the use of non-protein energy sources and stated that xylitol can be utilized both by man and animals, but emphasized that only the initial steps in xylitol metabolism are insulin-independent. As has been several times established in this survey, physicians and nutritionists to a certain extent disagree with regard to the value of fructose, sorbitol and xylitol in modern parenteral nutrition. One of the points raised by Lee and Wretlind [638] was that sorbitol, xylitol and fructose, although they are metabolized, cannot be directly used by peripheral muscle tissue before conversion in the liver to glucose. They further stated that glucose substitutes 'offer no metabolic advantages over glucose and have been associated with an impressive list of metabolic disturbances'. This seems to be, however, a somewhat narrow view, as indicated by the literature available describing the success in parenteral feeding of carbohydrate mixtures in special clinical conditions. These authors [638] listed a number of metabolic disturbances caused by the use of the above energy sources under certain specified circumstances. These aspects of parenteral nutrition have been discussed, for example, in chapters 7.23, 7.24, 7.3 and 7.4. The overwhelming majority of published studies and the success in the practical utilization of polyols, discussed in the above chapters, show that these glucose substitutes clearly have a place in parenteral nutrition.

Evaluation of the significance of polyols in parenteral alimentation should be performed far-sightedly. So, for example, Heller [639], still realizing the value of glucose in long-term parenteral nutrition, brought forth the fact that the situation is quite different in the immediate postoperative period. The utilization of glucose is disturbed and gluconeogenesis from amino acids is increased in stress situations. Other accompanying phenomena can be hyperglycemia, glucosuria and ketosis, and considerable amounts of K^+ ions are lost. According to Heller, a possible solution is the administration of fructose, sorbitol and xylitol as precursors of glucose. Heller emphasized on the fact that fructose, sorbitol and especially xylitol are superior to glucose in the postoperative state, due to their considerable nitrogen-sparing and antiketogenic effect. An important advantage of these three glucose substitutes in parenteral nutrition is their delayed conversation into glucose. This causes less strain on the regulatory systems of the body [639].

Xylitol was used as a partial source of calories together with glucose, fructose and sorbitol in studies which also investigated the effectiveness of amino acids in total parenteral alimentation [640–642]. No adverse effects were encountered and the carbohydrate solutions containing xylitol were considered effective in improving patients' nutritional status. Wang et al. [641] also stated that administration of a combination of glucose and fructose is better tolerated than giving a concentrated glucose solution alone as an energy source in total parenteral nutrition. Schreier and Porath [643] made it clear that although glucose is to be preferred before all other energy sources, the glucose tolerance in premature infants is often very low. The tolerance to glucose further decreases in hypoxia, acidosis, surgical trauma etc. Fructose and even galactose are assimilated in these subjects faster than glucose. Schreier and Porath concluded on the basis of published metabolic studies and their own experience with several thousand intravenous drips that 'it is ingenious to combine carbohydrates which are metabolized by different enzyme systems'. Consequently, these authors have used xylitol and sorbitol together with glucose, keeping the maximum infusion rate of the polyols at 0.1 g/(kg × h).

9.3 Chemical aspects

Kankare [644] investigated the mannitol-boric acid system, but the findings made in this study may be applicable to many other polyols as well. In addition to the previously assumed 1:1 and 2:1 mannitol-borate complexes, the existence of a 1:2 mannitol-borate complex and 2:1 and 1:2 uncharged mannitol-boric acid complexes must be assumed [644, 645]. In general, polyols form neutral complexes (a) with boric acid in acidic water solutions, whereas in basic conditions they form complexes with the borate anion as well (b):

$$\begin{matrix} -OH \\ -OH \end{matrix} + B(OH)_3 \overset{H^+}{\rightleftharpoons} \begin{matrix} -O \\ -O \end{matrix}\!\!>\!B-OH + 2\,H_2O \qquad (a)$$

$$\begin{matrix} -OH \\ -OH \end{matrix} + B(OH)_4^- \overset{OH^-}{\rightleftharpoons} \begin{matrix} -O \\ -O \end{matrix}\!\!>\!\bar{B}\!<\!\begin{matrix} OH \\ OH \end{matrix} + 2\,H_2O \qquad (b)$$

A new assay for polyols was developed by Bok and Demain [646]. The method was suggested to minimize interference by sugars. It was based on oxidation of alditols to formaldehyde by acidic periodate, reduction of the excess periodate with L-rhamnose, and a short-time determination of the formaldehyde with Nash reagent. Glycerol produced the strongest color reaction. Xylitol was not tested.

10. List of animal and human studies (excluding dental studies)

Table 8.
Animal and human* studies involving oral or parenteral administration of xylitol, and other studies related to the metabolism of xylitol in mammals.

Authors	Description of study	Findings
1960		
Tudisco (1960) [407]	Rat, oral feeding.	2–5 g/kg was not toxic. 16–20 g/kg was toxic and produced diarrhea.
1961		
*Kumahara, Feingold, Freeberg and Hiatt (1961) [408]	Man, normal subjects and subjects with pentosuria and pentosuria trait (5–25 g glucuronolactone orally).	Endogenous serum L-xylulose concentration increased above 1 mg% in pentosuric patients. The level increased several-fold following administration of glucuronolactone. The ketopentose was not normally present in serum in most pentosuric subjects, but injection of glucuronolactone increased it.
*Mellinghoff (1961) [410]	Man, diabetic, orally 16–20 g/day up to 7 days or 15–60 g/day for 1 day.	Xylitol was found to fit well to diabetic subjects. Peroral administration of xylitol did not affect blood sugar values.
Haydon (1961) [409]	Rat, incubation of liver slices of starved animals.	Xylitol and ribitol were antiketogenic in the liver slices of starved rats.
Kieckebuch, Griem and Lang (1961) [411]	Rat, 10% diet over 12 weeks; 30% diet over 5 weeks. Mouse, orally or intraperitoneally.	No pathological changes in rat livers, kidneys or hearts. No cataract formation. No changes in blood chemistry. No effect on weight gain and fertility. 50% xylitol in rat diet was lethal in 1–2 weeks. LD_{50} per os for mice was 25.7 g/kg and 22.2 g/kg as intraperitoneal injection (i.e. as for other common sugars).
1962		
*Bässler, Prellwitz, Unbehaun and Lang (1962) [412]	Man, healthy (orally 40 g/day for 1 day or 1 g intravenously).	No pathological changes were found. Xylitol was recommended for diabetics.
Griem and Lang (1962) [413]	Rat, intravenously (7.5 and 10% solutions).	No degenerative or inflammatory changes in organs.

Authors	Description of study	Findings
1963		
*Prellwitz and Bässler (1963) [418]	Man, non-diabetic and diabetic (5–8.75 g as injection in 1–1.5 min.)	Intravenously administered xylitol was rapidly metabolized in both test groups. No blood sugar elevation was found. No differences between the groups were detected.
Bässler and Dreiss (1963) [414]	Rat, alloxan diabetic (orally up to 2 g/day).	Xylitol was antiketogenic in 50–80% of cases (35% for sorbitol). Mannitol and xylose were not antiketogenic.
Bässler and Heesen (1963) [415]	Rat, normal and alloxan diabetic (infusion of 2.52–3.15 g xylitol in 60–90 min).	Xylitol, sorbitol and glucose were converted into liver glycogen to an equal extent in both test groups. Sorbitol and xylitol did not yield significant amounts of muscle glycogen.
Czok and Lang (1963) [416]	Mice, 0.25–1.0 g/kg in diet.	0.5 g or less per kg body weight did not have any effect on intestinal peristalsis.
Foglia, Yabo, Bernaldez and Aguirre (1963) [417]	Rats, normal and 95% pancreatectomized (gastric intubation and ad libitum, up to 1 month).	6% xylitol in the diet did not modify the course of developing diabetes caused by 95% pancreatectomy, as based on chemical and histological findings. No effect on fasting blood sugar and pancreatic histology. No adverse effects.
1964		
*Mehnert, Summa and Förster (1964) [419]	Man, diabetic or with liver disease (intravenously, 0.5–1.0 g/kg for 90 min).	0.5 g/kg: no significant differences between the subjects, for example, blood sugar levels in normal persons did not change. 1.0 g/kg: slight tendency for delayed utilization at higher age and liver disease.
Bässler and Prellwitz (1964) [423]	Rats, evisceration.	Insulin had no effect on the distribution of xylitol in eviscerated diabetic rats.
Müller (1964) [424]	Rats, histological study.	Xylitol was primarily absorbed at the apex of the intestinal villi. Absorbed xylitol (or its metabolized products) were distributed in the epithelial layer of the villi. Xylitol was then further transported into the connective tissue of the villi and finally into the vessels.
Schmidt, Fingerhut and Lang (1964) [420]	Rats, feeding or intravenously (alloxan diabetic and normal rats).	After adaptation the turnover rate of xylitol raised 3-fold in the feeding study both in alloxan diabetic and control rats. The maximal capacity to oxidize xylitol after intravenous injection was 50 ± 5 mg/(hr$\times$kg).
1965		
Bässler and Reimold (1965) [316]	Rat, erythrocyte suspensions.	Lactate was produced from glucose, fructose, xylitol, sorbitol and ribitol in erythrocytes. Mannitol had no effect. Combinations xylitol-glucose, xylitol-fructose and sorbitol-glucose had an additive effect, whereas combinations xylitol-sorbitol, xylitol-ribitol and glucose-fructose had no additive effect.

Authors	Description of study	Findings
*Manenti and Della Casa (1965) [421]	Man, diabetic, oral feeding (20 g/day).	No adverse effects.
*Yamagata, Goto, Ohneda, Anzai, Kawashima, Chiba, Maruhama and Yamauchi (1965) [389]	Man, diabetic and controls (intravenously, long- and short-term tests, 30 g/90 min).	Xylitol produced benefits to the metabolic disorders in diabetes mellitus. No deterioration of the diabetic state, rather, xylitol brought improvement in diabetic ketosis. No harmful effects. No changes in liver function tests. Xylitol diminished glycosuria.
*Bässler, Holzmann, Prellwitz and Stechert (1965) [426]	Man, intravenously (20 g).	Slight increase in ketopentose excretion in five out of nine cases was observed. This is far less than observed with similar load of xylose.
1966		
Hirata, Fujisawa, Sato, Asano and Katsuki (1966) [427]	Dogs, normal and diabetic (intravenously, 10% solution, 0.4 g/kg rapidly).	Blood sugar levels decreased significantly in normal dogs, the lowest level after 30 minutes. Plasma insulin increased significantly 10–20 minutes after injection. In diabetic dogs blood sugar levels increased, highest levels at 60 minutes, but no significant increase in plasma insulin.
Kuzuya, Kanazawa and Kosaka (1966) [385]	Dogs, intravenously (0.4 g/kg).	Marked rise in plasma insulin. After injection, plasma glucose increased slightly and then decreased below fasting level.
*Lang, Frey and Halmágyi (1966) [428]	1135 surgical patients, aged from 1 day to 80 years (intravenously, 5 and 10% solutions).	No side effects. Xylitol was recommended for intravenous nutrition in surgical patients because of its good compatibility.
Bässler, Stein and Belzer (1966) [429]	Rats, orally.	The rate of intestinal absorption of xylitol in rats adapted to xylitol was significantly faster than in control rats. There was no evidence of active transport. Adapted animals displayed higher rate of dehydrogenation and xylitol turnover than controls. Exogenous xylitol was dehydrogenated exclusively in the cytoplasm of liver cells. Other tissues were of little importance.
*Bässler, Toussaint and Stein (1966) [430]	Children (7 days to 8 years) and adults (30–41 years), intravenously 5–10% solution during 5–10 minutes.	There were no differences in the speed of the elimination process of xylitol between premature infants, mature infants, 4–8-year-old children, and adults aged 30–41 years. Premature infants possessed full capacity to use xylitol.

Authors	Description of study	Findings
1967		
*Toussaint, Roggenkamp and Bässler (1967) [431]	3–14-year-old children with acetonaemic reactions and diabetes mellitus (500 ml of a 10% solution in 3–4 hrs).	No influence on blood sugar levels. Xylitol had a good antiketogenic effect.
Müller, Strack, Kuhfahl and Dettmer (1967) [388]	Normal and alloxan diabetic rabbits (continuous infusion, 0.5 g/(kg × hr).	Xylitol was metabolized by both groups as well as fructose. 70% was utilized in liver. Xylitol was absorbed and excreted passively. Hepatic glucose production and plasma glucose were not affected by xylitol in controls, but increased in diabetic animals.
Montague, Howell and Taylor (1967) [422]	Rat, isolated islets of pancreas.	Xylitol, ribitol and ribose markedly stimulated the release of insulin. Ribitol and ribose were as effective as glucose, but xylitol was more effective. Both xylitol and glucose were effective through a common adrenaline-sensitive route.
Ross, Hems and Krebs (1967) [533]	Rat, liver perfusion.	Glucose was formed at a high rate from xylitol. A high activity of transketolase (EC 2.2.1.1) and transaldolase (EC 2.2.1.2) was postulated.
*Geser, Förster, Pröls and Mehnert (1967) [310]	Man, intravenously, as a 40% solution [0.8 g/(kg × 15 min)].	The slight increase of serum insulin was not statistically significant, the results thus differing from those of many animal experiments. Serum phosphate and free fatty acids were reduced after xylitol infusion. Blood glucose lowered first after 120 minutes.
*Akazawa, Iehara, Mori, Hashimoto and Tanaka (1967) [425]	Man, diabetes mellitus, chronic hepatitis, drug-induced hepatitis, surgery, diabetic retinopathy (parenterally).	Improvement in all states. Deterioration of diabetic state was avoided on surgical operation.
*Yoshikawa (1967) [432]	Man, non-diabetic and diabetic, in surgical operations [infusion, 0.4 g/(kg × hr)].	The concentration of blood xylitol, glucose, pyruvate and lactate behavied almost similarly in both groups. A patient with pancreatic carcinoma and severe hepatic dysfunction exhibited increased blood glucose and pyruvate.
1968		
*Klyachko, Perelygina and Domareva (1968) [433]	Man, with diabetes mellitus (intravenously).	No substantial rise in the level of glycemia. Xylitol was recommended as a sugar substitute and an antiketogenic agent in complex treatment of diabetes mellitus.
Podryadkov (1968) [434]	Dogs, ingestion.	Xylitol had a pronounced choleretic action, causing an augmented amount of bile, lipid complex, cholic acid, total phosphorus and bilirubin. Large doses inhibited external secretory function of liver.

Authors	Description of study	Findings
Hirata, Fujisawa, Sato, Asano and Katsuki (1968) [435]	Dog, normal and alloxanized, injection of 0.4 g/kg as a 10% solution.	Normal dogs: blood pyruvate decreased significantly, insulin increased more than after injection of glucose, accompanied by hypoglycemia. These responses were not found in diabetic dogs. Hypoglycemia might be caused by hypersecretion of endogenous insulin from β-cells.
Montague and Taylor (1968) [436]	Rat, isolated islets of pancreas.	The islets secreted insulin in response to xylitol, ribitol and ribose, but not to sorbitol, mannitol, arabitol, xylose or arabinose. 1 μM adrenaline inhibited the effect of glucose and xylitol on insulin release. Xylitol did not obviously stimulate insulin release by alterations in the intracellular concentrations of cAMP.
Paschmi and Opitz (1968) [437]	Rabbit, guinea-pig, mouse, rat (50% solution intravenously or 20% solution intraperitoneally).	Xylitol caused hypoglycemia and disturbances in the nervous system (see chapter 7.3).
Stein and Bässler (1968) [438]	Rat, infusion of 2 ml of 50% solution in 1 minute.	During infusions of fat emulsions (1.65 g in 1 hr), the elimination constants for xylitol were reduced less than for glucose. At a lower administration rate of fat, the elimination of xylitol was not impaired.
*Schultis and Geser (1968) [439]. See [440] (1970)	Man, postoperative state (intravenously 100 g/day).	Xylitol was recommended for source of energy during postoperative stages.
*Toussaint (1968) [441]	Children (3–14 years) with diabetes mellitus, 35 infusions, 1.5 g/kg.	The antiketogenic effect of xylitol was good. Xylitol was regarded as a useful source of energy in young diabetics.
*Geser and Mehnert (1968) [442]	Man, normal, diabetic or with liver disease.	The use of xylitol was considered advantageous in diabetics. The insulin-stimulating effect of xylitol, found in test animals, could not be encountered in man.
*Aono (1968) [443]	Man, healthy, intravenously (10 ml/kg of a 5% solution in 30 min).	Metabolic disturbances during ether anaesthesia were less severe in xylitol than in glucose infusion.
*Pitkänen and Sahlström (1968) [444]	Man, oral intake of 10 g glucuronolactone or 10 ml ethanol (or infusion of 500 ml of 10% ethanol).	Administration of either compound resulted in an increase in urinary xylitol. A previous study showed xylitol in normal human urine in low concentrations [446, 536].
*Halmágyi and Israng (1968) [445]	Man, infusion of 1189 patients (1497 infusions) during 1963–1965 (0.75 g/kg for 30 min).	The utilization of xylitol was not disturbed by the postoperative state. The use of xylitol in infusion therapy was recommended.

Authors	Description of study	Findings
*Fujisawa (1968) [447]	Dogs and man (intravenously).	Normal dogs: Blood sugar and pyruvate decreased, plasma insulin increased. Slight increase in lactate and α-ketoglutarate. Alloxanized diabetic dogs: Blood sugar, pyruvate and lactate elevated; no change in plasma insulin. Healthy volunteers: Slight increase in sugar and insulin. Decrease of pyruvate and slight increase of lactate. Diabetics: Sugar, insulin and pyruvate increased.
*Kondo, Naito and Sugiura (1968) [448]	Man, intravenously 50 g in 500 ml.	Urinary β-glucuronidase and glucaric acid decreased, but no relationship between these findings was suggested to occur.
*Iwasa and Takahashi (1968) [449]	Man, prevention of side effects of mitomysin C.	Xylitol alleviated side effects.
*Marimjan, Gusarenko, Lanshina, Popova and Tkatsenko (1968) [450]	Man, with diabetes mellitus (xylitol in confectionary items).	The use of xylitol in the diet of diabetic patients was recommended.
1969		
Ishii, Takahashi, Mamori and Murai (1969) [451]	Rats, treated with CCl_4 or alloxan.	Xylitol was found to be a good source of energy of rats treated with alloxan or CCl_4.
*Coats (1969) [452]	Man (a single patient for 7.5 months on xylitol infusion); 1278 bottles of infusion fluid was used.	The patient made complete recovery from the underlying disease (necrosis in small bowel and superior mesenteric artery thrombosis followed by operations) and was transferred to satisfactory oral feeding. Even at an infusion rate of 62.5 g per hr, the urinary losses were less than 4%.
Blockus, Donahoe, Crowley, Keating and Weinberg (1969) [453]	Monkeys and rabbits (infusion).	The authors indicated that they have developed adequate safety for xylitol in the treatment of diabetic patients.
Kuzuya and Kanazawa (1969) [454]	Dogs, intravenously [0.5–1.0 mg/(kg × min)].	Infusion of xylitol into the superior pancreatico-duodenal artery increased plasma insulin, suggesting that xylitol had direct stimulatory effect on islet cells. A common mechanism between the insulin-releasing effects of glucose and xylitol may exist.
Sheleketina (1969) [356]	Rats, xylitol in the diet.	Increase in liver glycogen, vitamins C and B_1 and niacin. This may explain the curative effect of xylitol in affections of the hepatobiliary system.

Authors	Description of study	Findings
*Dubach, Feiner and Forgó (1969) [455]	Man, healthy, oral administration (75 g/day for 21 days; also up to 220 g/1 day).	No diarrhea at less than 130 g/day. At 220 g/day aversion to sweets. Weight and fasting blood sugar normal. Xylitol caused less flatulence and meteorism than sorbitol and was preferred by the subjects.
*Kinami and Kitagawa (1969) [456]	Man, surgical operations.	Xylitol was found to be a suitable source of energy in the cases described.
Kuzuya, Kanazawa and Kosaka (1969) [457]	Dog, injection (0.4 g/kg) or orally (1.0 g/kg).	Xylitol, glucose, fructose and sorbitol (in this decreasing order) increased plasma insulin significantly. Mannitol did not increase. Xylitol and fructose increased plasma glucose slightly, then decreased. Xylitol and glucose (0.05 g/kg) produced comparable effects, but the effect of xylitol became more pronounced with increasing dosage. Oral administration of 1.0 g xylitol per kg also produced greater hyperinsulinemia than glucose. Xylitol augmented the secretion of insulin from the pancreas.
*Hirata, Fujisawa and Ogushi (1969) [458]	Dog, normal (1.0 g/kg). Man, normal (0.5 g/kg). Both intravenously.	The hyperinsulinemic effect of xylitol in normal dogs was completely suppressed by epinephrine. In man the effect of xylitol on plasma insulin was much less than in dog and individual differences were observed.
Hosoya and Machiya (1969) [459]	Rat liver slices of alloxan diabetic animals, ^{14}C-palmitate as substrate.	Xylitol did not affect the complete oxidation of palmitate. Xylitol would provide energy for β-oxidation of palmitate. It is likely that xylitol decreases accumulation of ketone bodies.
*Erdmann (1969) [460]	Children, rabbit (infusion).	Xylitol and sorbitol were regarded as important sources of energy in parenteral nutrition of infants, as well as in the treatment of ketosis due to diabetes.
*Yamagata, Goto, Ohneda, Anzai, Kawashima, Kikuchi, Chiba, Maruhama, Yamauchi and Toyota (1969) [461]	Man, normal and diabetic (intravenously 30 g in 90 min or orally 30 g).	Xylitol was considered an important tool in the treatment of diabetic ketosis and as a parenteral source of energy. Xylitol could be used as a substitute for sucrose, which does not effect plasma lipids or mucoprotein.
*Mehnert, Förster and Dehmel (1969) [462]	Man, normal, diabetic or with liver disease (intravenously 0.5 g/kg in 90 min or 1.0 g/kg in 30 min).	Only slight differences in the utilization of xylitol between the groups. Older subjects and patients with liver disease had a lower utilization of xylitol, as is the case with all carbohydrates. In diabetics blood glucose did not increase as happens after administration of glucose or fructose.

Authors	Description of study	Findings
*Uesuka, Tosen, Ohba, Akamatsu, Narahara and Kodaira (1969) [463]	Man, with partial gastrectomy, intravenously 75 g xylitol/day for 6 days (or 10% xylitol in 120 min).	Administration of xylitol to postoperative diabetic patients as the caloric source is preferable on accountof its slight increasing effect on blood sugar and its antiketogenic action. No remarkable difference between xylitol and glucose as regards fasting blood sugar level.
*Ishii and Sambe (1969) [464]	Man, normal or with diabetes mellitus or liver cirrhosis. Rats (CCl_4-poisoned or alloxanized). Man: 25 g/60 or 90 minutes. Rat: in vitro with liver homogenates.	Xylitol level decreased rapidly after administration, with small urinary loss of xylitol. No marked elevation of blood glucose or pyruvate. Rats: labelled xylitol was easily incorporated into liver glycogen and oxidized to CO_2 both in CCl_4-poisoned and alloxanized rats.
*Schultis and Geser (1969) [465]	Man, after laparatomy or with fractures, infusion, 1.5 g/(kg $\times$ hr).	A normalization of the disturbances of glucose utilization was found in operated patients, as well as reduction in ketone bodies. Obviously a reduced loss of nitrogen. Compared with sorbitol and fructose, xylitol utilization showed some advantages. Serum insulin was not affected by xylitol in healthy persons.
Opitz (1969) [466]	Rat, adipose tissue, in vitro, and rat, in vivo (fat mobilization).	Xylitol caused a significant drop in plasma free fatty acids without any concomitant change in blood sugar concentration.
*Dehmel, Förster and Mehnert (1969) [467]	Man, rat (injection into duodenum or gastric intubation).	Xylitol was absorbed very slowly. Therefore, it could be useful in the diet of diabetic persons. 40 g/day was recommended for the upper level for diabetic subjects in this study.
Ohta, Takano and Kosaka (1969) [468]	Rat, orally (20–40% of all carbohydrate).	10 or 20% xylitol in a choline-deficient diet modified the production of fatty liver and fatty liver cirrhosis of rat. Xylitol prevented hemorrhagic renal necrosis.
Wang and Meng (1969) [469]	Rat, minced lung.	Xylitol was utilized and metabolized by lung tissue.
Lambert, Junod, Stauffacher, Jeanrenaud and Renold (1969) [470]	Organ culture of fetal rat pancreas.	11 mM xylitol, ribitol and D-ribose were ineffective in causing insulin release in the presence of 10 mM caffeine, whereas glucose, D-mannose, D-fructose and D-galactose (in this decreasing order) were effective.
Hosoya and Iitoyo (1969) [471]	Rats, oral feeding (up to 20% xylitol in diet for 4 months).	Transient dose-dependent diarrhea. When rats adapted to xylitol, their hepatic NAD-xylitol dehydrogenase activity was induced. No effect on pregnancy, birth performance and growth of offspring. The offspring needed the same adaptation as that observed with parent rats.

Authors	Description of study	Findings
*Wolf, Queisser and Beck (1969) [472]	Man, patients with liver cirrhosis, fatty liver, and controls. Infusion of 50 ml of a 40% solution of glucose, fructose, sorbitol and xylitol.	Fructose, sorbitol and xylitol caused a drop in serum P_i, which was more marked than with glucose. The glucose effects were similar in all groups, whereas the effects of the other sugars on P_i levels declined with increasing severity of liver damage. It was explained that fructose, sorbitol and xylitol are rapidly metabolized by the liver in contrast to glucose which is used peripherally to a higher degree, but more slowly.
*Berg, Leis and Ohnhaus (1969) [473]	Man, with chronic uremia (slow injection of a 40% solution).	Xylitol was utilized in uremic patients.
*Maynard (1969) [534]	Man, intravenous infusion with a 50% solution (1 litre in 8 hrs).	Xylitol was tolerated better than sorbitol. Xylitol was considered to have three routes to enter the glycolytic pathway (one with sorbitol).
1970		
*Donahoe and Powers (1970) [264]	Man, intravenously, 5 or 10% solutions, 1.22–3.13 g/kg.	Xylitol administration produced hyperuricemia, hyperuricosuria and hyperbilirubinemia (see chapter 7.3).
*Spitz and Rubenstein (1970) [474]	Man, uremic and controls (infusion, 0.33 g/kg of a 50% solution over 2 min).	Xylitol was well utilized in renal failure. It is rapidly cleared from the blood, and urinary excretion is minimal. Xylitol induced a marked fall of plasma P_i, indicating peripheral utilization. Xylitol was considered a useful source of calories in uremia and other conditions characterized by carbohydrate intolerance and insulin resistance.
Landgraf and Matschinsky (1970) [475]	Mouse, rapid injection.	The differential permeability of β-cells and acini to xylitol is similar to that of glucose and sorbitol, as found also earlier [476]. However, xylitol and sorbitol enter β-cells more slowly than glucose. The study showed that insulin-releasing sugars and polyols have a selected permeability for β-cells.
Shima, Mitsunaga and Nakao (1970) [379]	Rat, testicular homogenates incubated in the presence of xylitol or glucose.	Both sugars increased the incorporation of precursors to testicular protein and RNA. Both sugars increased ATP levels. Both sugars were considered similar as regards the mechanism involved.
*Schultis and Geser (1970) [440]. See [439] (1968)	Man, postoperatively (intravenously 100 g/day over 8–20 hrs).	It was considered that the metabolic disturbances caused by stress can be influenced favourably by xylitol, while glucose has practically no influence on these disturbances.

Authors	Description of study	Findings
*Spitz, Bersohn, Rubenstein and van As (1970) [477]	Man, healthy, infusion of 500 ml of a 10% solution for 60 minutes.	Xylitol (or its metabolite) increased the amount of growth hormone in serum more than glucose (infusion of fructose may suppress the growth hormone level [482,483]). However, Geser [578] added that growth hormone secretion is depressed by glucose, fructose and sugar alcohols, but in pathological conditions fructose and xylitol are capable of stimulating the release of growth hormone.
Geser, Schultis and Diedrichson (1970) [478]	Dog, injection.	Extremely high increase in blood insulin. In man such an increase is obtained only after extremely high doses of xylitol.
Swarm and Banziger (1970) [479]	Rat, oral feeding, 5–20 g/(kg × day). Analyses during 4–13 weeks.	All animals survived. Some reduced weight gains were dose-dependent. Transient diarrhea, but otherwise no abnormalities in ophthalmic, neurologic, hematological, blood and urine examinations. Treated animals could not be distinguished from controls.
Banziger (1970) [480]	Monkey *(Macaca rhesus)*, oral feeding, 1.0–5.0 g/(kg × day) for 13 weeks.	Transient dose-dependent diarrhea. No changes compared to controls in behaviour, appetite, weight, ophthalmoscopic and neurologic examinations, clinical laboratory studies, organ weights, and gross and microscopic pathology.
Wilson and Martin (1970) [481]	Dog, rhesus monkey, intravenously.	Xylitol, glucose and tolbutamide increased plasma insulin significantly, xylitol being most effective in dogs and least in monkeys. Xylitol and glucose may stimulate insulin release through a common metabolite of the pentose phosphate pathway.
*Thomas, Edwards and Edwards (1970) [263]	Man, intravenously.	A number of severe side effects were reported. The authors indicated that the xylitol solutions may have contained a contaminant (see chapter 7.3).
*Spitz, Rubenstein, Bersohn and Bässler (1970) [384]	Man, uremic and controls, intravenously [500 ml of a 10% solution in 60 min, or 0.33 g/(kg × 2 min)].	Xylitol stimulated the release of insulin, but the levels were lower than after glucose. Xylitol was recommended as a source of energy in uremia and in conditions of carbohydrate intolerance and insulin resistance.
Jacob, Asakura and Williamson (1970) [484]	Rat, liver perfusion.	Xylitol may cause maximal stimulation of glucose production by increasing triose phosphate levels in tissues. Xylitol decreased ketone body production in the presence of exogenous fatty acids (possibly by increasing esterification of fatty acids by α-glycerophosphate). Xylitol decreased the ATP/ADP ratio, possibly by decreasing the flux of the citric acid cycle.

Authors	Description of study	Findings
Matschinsky, Ellerman, Kotler-Brajtburg, Krzanowski, Fertel and Landgraf (1970) [485]	Rat, intravenous administration.	Both glucose and xylitol increased insulin levels promptly in peripheral blood already within 1 minute.
*Mehnert, Förster, Geser, Haslbeck and Dehmel (1970) [364]	Man, normal, diabetic and subjects with liver disease (0.5 g/kg); parenteral nutrition.	With patients who have impared glucose tolerance, insulin-independent sugars or polyols should be used instead of glucose for parenteral nutrition. Fructose, sorbitol and xylitol do not alter the glucose concentration in blood, with the one exception of ketoacidotic diabetes. Xylitol stimulates insulin secretion in unphysiological high dosages only.
*Förster, Meyer and Ziege (1970) [540]	Man, healthy, infusion, 1.5 g/kg in 20–25 minutes as a 20% solution. Rat, infusion.	Serum urate and bilirubin were transiently raised after sorbitol, fructose and xylitol. No change in transaminases. In two subjects of three given glucose the same raise in bilirubin was found as after the other sugars. The increase of urate and bilirubin after rapid infusion of the sugars are not toxic effects. Rat: 20 g/(kg × day), no effect on bilirubin or transaminases.
1971		
*Grabner, Berg, Bergner, Matzkies and Maeder (1971) [486]	Man, diabetic, intravenously (500 ml of a 10% solution in 90 min).	Xylitol reduced serum acetoacetate, β-hydroxybutyrate, unesterified fatty acids, free glycerol and total lipids, when compared to control experiments. Hyperlipidemia might be favourably influenced by xylitol.
*Wang, Patterson and van Eys (1971) [487]	Man, normal or with glucose 6-phosphate dehydrogenase deficiency (incubation of erythrocytes). Rabbit, intravenously, 0.5 g/kg for 6 hours.	Injections of xylitol to rabbits were considered non-toxic. In vivo and in vitro experiments showed that xylitol partially prevented acetylphenylhydrazine-induced acute hemolysis.
*Turner, Schneeloch and Nabarro (1971) [488]	Man, normal, diabetic and obese. Intravenously, 0.5 g glucose per kg or 0.383 g xylitol per kg for 2 minutes.	Normal and obese: glucose was more effective than xylitol in stimulating the 1st phase of insulin release. Diabetics: the initial response to xylitol was not reduced to the same extent as the response to glucose. Xylitol and glucose equally stimulated the 2nd phase in all groups, but xylitol gave a less prolonged response. Neither xylitol nor glucose stimulated either phase in severely diabetic subjects. Both sugars may have a common stimulatory mechanism (via pentose phosphate pathway).

Authors	Description of study	Findings
Pool (1971) [489]	Dog, oral feeding, 10 g/(kg × day).	No change in pH or P_{CO_2} in blood.
Mosinger (1971) [490]	Rat, oral feeding 0.1 g/(kg × day) for 11–24 months.	No effect on reproduction (three generations). No malignant tumor development. No other pathological changes in the rats.
Williamson, Jacob and Scholz (1971) [491]	Rat, liver perfusion.	In the experimental system described addition of xylitol increased the rate of glucose formation and decreased the rate of ketogenesis. Xylitol also increased α-glycerophosphate levels and induced a depletion of P-enolpyruvate and 3-glycerophosphate in substrate-depleted livers.
Froesch, Zapf, Keller and Oelz (1971) [291]	Rat, normal and streptozotocin-diabetic (intravenously with labelled compounds).	Young male rats metabolized xylitol, sorbitol and fructose with a half-life of 165 seconds. Within 10 minutes more than 70% of ^{14}C in plasma was as ^{14}C-glucose. The labelling of glucose was insulin-independent. A small percentage of label was retained by liver as glycogen and lipids in the normal rat. In diabetic rat insulin enhanced slightly glycogen synthesis from xylitol and reduced its incorporation into lipids.
Keller and Froesch (1971) [492]	Rat, fasted and streptozotocin-diabetic (intravenously with labelled xylitol).	The similarities of the metabolism of glucose, xylitol and fructose were considered to be due to the rapid conversion of the two latter to glucose.
*Kuzuya, Kanazawa, Hayashi, Kikuchi and Ide (1971) [493]	Man, cow, horse, goat, rabbit, rat (intravenously).	Marked species differences were found. In man 12–30 g glucose produced greater rises of plasma insulin than xylitol. The results in other species were in contrast to those obtained with dogs in which xylitol is a stronger insulin-releaser than glucose.
*Amador and Eisenstein (1971) [494]	Man, oral feeding up to 20 g/day, or 20–40 g/m² body surface.	Slight and rapidly reversible elevation of plasma urate and lactate at high dose levels. One subject had diarrhea and flatus at high dose levels. No effect on transaminases, lactate dehydrogenase and alkaline phosphatase. High doses slightly increased plasma glucose, insulin and bilirubin. Urate and bilirubin were normal in 8 hours. No gastrointestinal distress at less than 90 g/day.
*Humke and Wiessmann (1971) [495]	Pregnant women, infusion of a 10% solution (500 ml in half-isotonic electrolyte solution).	The high keto-body concentrations measured during tedious labour can drop by 500 ml solution of 10% xylitol (in half-isotonic electrolyte solution) to results displayed by labours of normal duration.

Authors	Description of study	Findings
*Schumer (1971) [298]	Man, two diabetics and two non-diabetics (postoperatively, infusion of a 10% solution, 1.5 g/kg). Two non-diabetics loaded with 4.5 g/kg.	1.5 g/kg: increase of lactate, urate, bilirubin and alkaline phosphatase. 4.5 g/kg: pain in the right upper quadrant of the abdomen, vertigo, headache, nausea, vomiting, all on the first day after infusion. Serum and urine urate and serum bilirubin, aspartate aminotransferase, lactate, alkaline phosphatase and P_i were significantly increased, but returned in 10 days to normal values after cessation of infusion (see chapter 7.3).
1972		
*Heuckenkamp and Zöllner (1972) [496]	Man, healthy, intravenously 0.5 g/(kg × hr) for 5 hours.	Xylitol levels in plasma increased up to 80 mg/100 ml. Xylitol excretion in urine averaged ca. 22% of the dose. Within 165 minutes serum urate rose from a fasting value of 4.7 mg/100 ml to 7.2 mg/100 ml. No controls were used.
*Förster (1972) [365]	Man, healthy, intravenously, 1.5 g/kg as a 20% solution in 20–25 minutes. Rat, intravenously, 20% solution for 68–72 hours at a rate of 1.2 ml/hr (i.e. 18 g/kg sugar in 24 hrs).	Man: rapid infusion of fructose, xylitol or sorbitol led to a significant increase of serum urate. This was not found with glucose or galactose. Fructose, sorbitol, xylitol and glucose raised serum bilirubin. Both transaminases were unaltered. Lactate increased after glucose, fructose and sorbitol, but not after xylitol. Rat: fall in serum bilirubin and transaminases after fructose, sorbitol and xylitol. The raise of urate was considered to be due to increased purine synthesis. Also sucrose causes raises in serum urate. The author emphasized the normality of these reactions and recalled several weaknesses in the studies of Schumer [298] and Thomas et al. [265].
*Willgerodt, Beyreiss and Theile (1972) [497]	Mature and premature newborns, intravenously 0.5–1.0 g/kg rapidly (2–3 min) as a 20% solution. Orally: 0.5–1.0 g/kg as a 20% solution.	Blood glucose increased significantly 5 minutes after infusion and remained high even after 120 minutes. Rapid infusion also increased lactate, but standard hydrogen carbonate and base-excess did not change. Xylitol should not be infused rapidly with newborns. Excretion of xylitol in urine: 8.4–15.5% of dose. Oral use did not affect above serum parameters.
Staub and Thiessen (1972) [498]	Rat, polyols in hypercholesterolemic diet (27 or 54% of total diet).	54% diet: after 35 days 87% of sorbitol- and 93% of mannitol-eating died, whereas two thirds of those eating xylitol survived.Rats fed sorbitol or mannitol exhibited higher cholesterol levels than those fed xylitol. The study suggested that these polyols are more hypercholesterolemic than starch or sucrose.

Authors	Description of study	Findings
*Mertz, Kaiser, Klöpfer-Zaar and Beisbarth (1972) [499]	Man, healthy, intravenously (10% solution for 100 min; 5 ml/min).	The concentration of serum total lipids, triglycerides, glycerol, α-lipoprotein, total cholesterol, phosphatides, acetoacetate and β-hydroxybutyrate were not changed significantly. Serum free glycerol raised immediately after infusion. Part of released glycerol was considered to have raised from xylitol. At the end of infusion a slight lactate increase, but not in pyruvate. Serum urate was increased 30%. De novo synthesis of urate may be due to enhanced nucleic acid turnover in liver.
*Mertz, Kaiser, Klöpfer-Zaar and Beisbarth (1972) [500]	Man, orally 15–50 g/day for 2 weeks.	No side effects. No significant changes in the following serum parameters: triglycerides, non-esterified fatty acids, free cholesterol, α-lipoproteins, β-lipoproteins, total cholesterol, phosphatides, acetoacetate and β-hydroxybutyrate. Serum P_i rose until the 7th day and remained then steady and slowly decreased thereafter. Lactate levels declined from beginning. Serum urate was unchanged. Metabolic effects of xylitol in healthy subjects were considered favourable.
Wang, King, Patterson and van Eys (1972) [501]	Rabbit, intravenously, 0.5–0.8 g/kg.	Dose of 4 g/kg as a 50% solution: transaminases, creatine phosphokinase, lactate dehydrogenase increased sharply. No change in urate, cholesterol, P_i, glucose or alkaline phosphatase levels. Minor elevations of serum bilirubin in some animals. No increase in urine urate and oxalate. In 5% solutions up to 2 g/kg xylitol was essentially non-toxic in the rabbit.
Brinkrolf and Bässler (1972) [377]	Rat, infusion of 1 mmole of xylitol or a mixture of fructose, glucose and xylitol (0.3 mmole each); 1.5 g/(kg × hr).	Large doses of fructose, sorbitol, or a mixture of glucose, fructose and xylitol caused a drop of liver ATP, total adenine nucleotides and P_i, and a rise in AMP. These changes were considered a transient disturbance of homeostasis by compounds which are rapidly phosphorylated in the liver. Doses of fructose, xylitol or the other carbohydrates not exceeding 0.5 g/(kg × hr) are not likely to disturb liver metabolism by changing levels of adenine nucleotides.
Jourdan, Macdonald and Henderson (1972) [502]	Baboon, intravenously, xylitol-U^{14}C.	Incorporation of label into serum glycerides was of the same order as found for fructose. Xylitol increased serum insulin and reduced serum glucose. Xylitol was converted into serum glycerides via the same pathways as glucose. The kinetics of the pathway was not found to be sex-dependent.

Authors	Description of study	Findings
Petrich, Reinauer and Hollmann (1972) [503]	Rat, various organ homogenates, Morris hepatoma and fibrocyte cultures.	The activity of the enzymes of the glucuronate-xylulose cycle was higher in epididymal adipose tissue (with the exception of NADP-xylitol dehydrogenase) than in the liver (based on soluble proteins). In brown adipose tissue and liver the activities were of similar magnitude. In Morris hepatoma the activity of the pathway was lower, but still comparable to that in liver. The enzymes were not detected in fibrocyte cultures. The high activity in the adipose tissue explains the extra-hepatic utilization of xylitol in normal and diabetic animals, and its lipolytic activity in adipose tissue.
*Thomas, Edwards, Gilligan, Lawrence and Edwards (1972) [265]	Man, intravenous infusion of patients with different diseases. Average rate: 0.23 ± 0.02 to 0.49 ± 0.03 g/(kg × hr). Total dose: 485 ± 83 to 1098 ± 305 g.	Ten of twenty-two patients were thought to have developed a syndrome of adverse reactions produced by xylitol solutions (diuresis, oliguria, azotaemia, liver and cerebral disturbances, etc.). Five patients were considered not to have shown any adverse reactions (see chapter 7.3).
*Thomas, Gilligan, Edwards and Edwards (1972) [347]	Man, one patient was given xylitol intravenously [250 g of a 50% solution in 6 hrs; 0.65 g/(kg × hr)], during and after treatment of tetanus.	Osmotic diuresis and lactic acidosis resulted from the use of xylitol in the way indicated. It was suggested that an increase in the lactate/pyruvate ratio results from infusion of xylitol (see chapter 7.3).
Mattson and Nolen (1972) [647]	Rat, oral feeding a fat diet which also contained xylitol pentaoleate.	Fatty acids of compounds containing less than 4 ester groups (methyl oleate, ethylene glycol dioleate, glycerol trioleate, and sucrose monooleate) were almost completely absorbed. The absorbability decreased with increasing number of ester groups: the absorbability of xylitol pentaoleate was 50%; sorbitol hexaoleate and sucrose octaoleate were not absorbed. The differences may depend on the activity and specificity of lipolytic enzymes in the lumen of the intestinal tract.
Mattson and Volpenhein (1972) [648]	Rat, feeding by stomach tube an emulsion diet in which the fat consisted of 95 parts of glycerol trioleate and 5 parts of one of several labelled polyol oleates, including xylitol 1-^{14}C-pentaoleate. Thoracic duct lymph was collected.	The percentage of the fed label which was detected in the lymph in 24 h was 88% from glycerol 1-^{14}C-trioleate, 67% from erythritol 1-^{14}C-tetraoleate, 24% from xylitol ester, and 2% from sucrose 1-^{14}C-octaoleate. Fatty acids esterified with glycerol appeared in the lymph most rapidly, while the transfer of the fatty acids of erythritol and xylitol into the lymph was slower. The amounts and type of appearance of fatty acids of the various esters were suggested to be related to the enzymes that hydrolyze these fats in the intestinal tract.

Authors	Description of study	Findings
1973		
Woods and Krebs (1973) [375]	Rat, liver perfusion.	α-Glycerophosphate accumulated, adenine nucleotides and P_i were depleted. Xylitol was rapidly metabolized, the main products being glucose, lactate and pyruvate. The loss of adenine nucleotides was considered to be due to depletion of P_i and accumulation of α-glycerophosphate, and not to increased ATP consumption. Abnormalities found by Schumer [298] and others were explained by increased degradation of hepatic adenine nucleotides. Intravenous infusion of relatively large amounts of xylitol is obviously not without risk (see chapter 7.3).
*Berger, Göschke, Moppert and Künzli (1973) [504]	Man, during abdominal surgery (infusion, 0.75 g/kg in 60 min, 6–8 days after surgery).	Immunoreactive insulin raised significantly after xylitol in portal vein, but not in peripheral vein. Blood glucose levels were unchanged. It was concluded that xylitol per se produces the secretion of insulin in vivo and in vitro.
*Asano, Levitt and Goetz (1973) [505]. See also [396]	Man, healthy, gastric intubation (5–10 or 15–30 g daily up to 2–3 weeks).	Xylitol absorption ranged from 49 to 95%. Chronic administration of 30 g/day for 2–3 weeks did not alter the rate of xylitol absorption. No abnormalities were found. Blood chemistry values remained normal.
*Dubach, Lejeune, Forgó and Bückert (1973) [506]	Man, orally in various forms for 14 days (10–80 g/day).	Abdominal distension in four cases out of 12 and diarrhea (five cases). Xylitol did not have any detectable effect on the number of the most important stool bacteria. Out of 15 bacterial species, xylitol inhibited only one lactobacillus at a concentration of 2–4%.
*Büngener (1973) [507]	Man, high-dose infusion therapy.	Reno-cerebral oxalosis was claimed to be a result of the use of xylitol in the way indicated (see chapter 7.3).
Blair, Cook and Lardy (1973) [508]	Rat, liver perfusion.	Glucagon stimulated glucose formation from xylitol and dihydroxyacetone. Formation of lactate from xylitol and oxidation of xylitol to CO_2 were inhibited 80 and 70%, respectively, by glucagon. It was suggested that the effect of glucagon on gluconeogenesis and glycolysis is the result of a complex interaction rather than regulation of one particular enzyme.
Birnesser, Reinauer and Hollmann (1973) [368]	Guinea-pig, rat. No administration of xylitol.	The activity pattern of pentulokinases of several guinea-pig organs (liver, kidney, heart, adipose tissue, spleen, muscle, brain, testis) was similar. In diabetic rats the activity of pentulokinases is diminished.

Authors	Description of study	Findings
Förster, Hoos and Lerche (1973) [509]	Rat, liver perfusion.	40–55% of added fructose or sorbitol was transformed into glucose, but 60–75% of added xylitol. The increase of hepatic lactate was higher with fructose or sorbitol than with xylitol. Application of xylitol, fructose or sorbitol did not affect the change for worse in carbohydrate metabolism, although extensive conversion into glucose takes place in the liver.
*Bickel, Bünte, Coats, Misch, von Rauffer, Scranowitz and Wopfner (1973) [510]	Man, after gastrointestinal operations [infusion, 0.48 g/(kg×hr)].	Xylitol was better utilized by patients in the postoperative period than by healthy controls. In patients with severe and persistant abnormality of glucose utilization, the capacity for xylitol utilization was also decreased.
*Matzkies, Grabner, Scharrer, Bickel and Berg (1973) [511]	Man, healthy, intravenously (0.125–0.5 g/kg for 6 hrs).	Blood glucose decreased during infusion. An insignificantly small insulin secretion occurred only after infusion of 0.5 and 0.375 g/kg in 6 hours. The direct effect of xylitol on β-cells cannot alone explain the finding (as suggested by Montague and Taylor [422, 436]. It is possible that xylitol inhibits gluconeogenesis in the same way as fructose and ethanol.
*Berg, Matzkies and Bergner (1973) [512]	Man, healthy, intravenously, 0.125–0.5 g/(kg×hr) for 6 hours.	Serum triglycerides were not changed significantly. Xylitol and fructose caused no effect on cholesterol concentration in serum, whereas sorbitol caused a significant decrease ($p<0.001$, although drop was only from 186 to 167 mg%).
Igarashi, Kobayashi, Shoji, Tanabe, Nomura and Ohtake (1973) [513]	Rabbit (fasted), infusion, 5% xylitol solution, 0.4 ml/min for 10 days.	Xylitol had sparing effects for nitrogen, water, Na and K, particularly at the late state of starvation (with glucose at early stage of starvation).
*Berg, Matzkies and Renz (1973) [514]	Man, infusion of 231 patients during 1969–1971 (15–20 g/hr over 6 days; total amount 50–7800 g).	Xylitol and fructose increased serum urate, but the values began to decrease during treatment. Both sugars increased bilirubin, but increases were found with other sugars as well. Urea nitrogen increased after all sugars tested (including glucose). Forty-seven cases of hypoproteinemia were observed, when most reductions in total proteins took place after xylitol. The sugar effects were dose-dependent. This report was examined by Heuckenkamp [541] who indicated that a dose of 0.5 g/(kg×hr) of fructose does not change urate, but administration of 1.0–1.5 g/(kg×hr) causes a noticeable increase. Xylitol, at a certain rate of administration, was claimed to cause a higher urate increase than a three times higher dosage of fructose [541].

Authors	Description of study	Findings
*Wenzel (1973) [515]	Man, infusion (500 ml of a 10% solution).	Benefits of xylitol and other sugars were discussed. Infusion with any carbohydrate was considered to require attention, because carbohydrate solutions are often very acidic and do not have significant buffering capacity.
*Berg (1973) [542]	Man, intravenously.	The author stated in this discussion that infusion of xylitol at a rate of 0.125–0.250 g/(kg × hr) leads only to a small or to no increase of serum urate.
*Berg, Bickel and Matzkies (1973) [381]	Man, healthy, parenterally up to 0.25 g/(kg × hr).	No change in lactate, pyruvate and HCO_3^- ions. 1:1 combination of fructose and xylitol and 2:1:1 combination of fructose, glucose and xylitol produced no side effects at a rate of 0.5 g/(kg × hr).
1974		
*Schröder, Féaux de Lacroix, Franzen, Klein and Müller (1974) [516]	Man, high-dose infusion therapy with xylitol, maximum 2.7–8.7 g/kg.	Reno-cerebral oxalosis was claimed to be a result of the use of xylitol as indicated (see chapter 7.3).
Grabner, Pesch, Matzkies, Bickel and Berg (1974) [517]	Rabbit, infusion of 50 ml of a 20% solution daily over 14 days [0.6 g/(kg × hr), for 6 hrs.].	The sole intake of xylitol (and sorbitol) did not cause morphological or functional disturbances of the adrenal cortex. The same concerned animals receiving xylitol and dexamethazone. Dexamethazone-induced adrenocortical insufficiency could be prevented by the simultaneous administration of xylitol in rabbits.
*Matzkies and Berg (1974) [518]	Man, healthy, infusion over 6 hours, 0.125–0.75 g/(kg × hr).	The concentration of neutral fats and cholesterol was not changed in blood, whereas total lipids reduced significantly.
*Kaiser, Stocker and Heuckenkamp (1974) [519]	Man, healthy, infusion [0.3–0.5 g/(kg × hr)].	Serum urate increased significantly. The authors stated that xylitol infusion leads to increased purine catabolism, because the concentration of 5-phosphoribosyl-1-pyrophosphate in erythrocytes was reduced.
*Heuckenkamp (1974) [520]	Man, infusion.	In many cases serum urate concentration was high at 24 hours after infusion. It was considered unlikely that a breakdown of preformed adenosine nucleotides would be the sole cause for the xylitol-induced hyperuricemia.

Authors	Description of study	Findings
*Berg, Matzkies and Bickel (1974) [521]	Man, healthy (infusion).	Dose limits for glucose: up to 0.75 g/(kg × hr) over a period of 12 hours. Raising the dose led to hyperglycemia. For sorbitol and fructose: up to 0.25 g/(kg × hr) over a period of 12 hours. A safe total dose of both was 210 g. Xylitol: up to 0.25 g/(kg × hr) (max. infusion time was 9 hrs). 0.125 g/(kg × hr) may be infused for 12 hours. The safe total dose was 158 g. Mixed solutions of fructose, glucose and xylitol can be infused in a dose of 0.5 g/(kg × hr) over 12 hours, the total dose being 420 g, i.e. 35 g/(70 kg × hr).
*Förster and Zagel (1974) [522]	Man, healthy, high-dose infusion, 1.0 g/(kg × hr) over 4 hours.	Glucose infusions may be followed by exaggerated counterregulations. They also led to an increase of lactate and glucose in serum. Free fatty acids increased more than 300% from the initial values and remained high over 4 hours. These reactions were rarely seen after infusion of xylitol and fructose. Because the two latter have a direct and insulin-independent action, these sugars may be advantageous in 'stress patients'.
*Wolf, Melichar, von Berg and Kerstan (1974) [523]	Man, premature infants, complete intravenous alimentation using a fat emulsion with amino acids, 10% xylitol and other ingredients (a caloric intake of 50–80 cal/(kg × day).	No complications were observed. The authors stated that long-term experiments in newborns and prematures have not been carried out (the study of Bässler et al. [430] was an elimination study).
*Geser, Müller-Hess and Felber (1974) [524]	Man, healthy, infusion of carbohydrate and amino acid mixtures (containing 50 g xylitol and sorbitol per liter).	Amino acids, in parenteral nutrition, should be given with carbohydrates to produce an anabolic effect. Insulin/glucagon ratio increased after infusion of sorbitol-xylitol-amino acid mixture, when compared to amino acids alone, but glucose + amino acids produced the strongest increase. In normal subjects the combination of amino acids and glucose has a strong effect on glucose storage and protein synthesis, but in patients this increase will not have the same meaning, due to frequent insulin insensitivity in surgical and burned patients. Sorbitol and xylitol potentiated the effect of amino acids on growth hormone secretion. The authors also stated that fructose, sorbitol and xylitol stimulate growth hormone secretion during surgical stress. This partly contradicts the suggestion that fructose may suppress growth hormone secretion [482, 483, 525].

Authors	Description of study	Findings
*Matzkies (1974) [526]	Man, intravenously, 0.125–0.75 g/(kg × hr) over 6 hours.	Serum Na, K and Ca decreased within the physiological range. P_i increased after fructose and sorbitol, but not after xylitol. Glucose concentration decreased after xylitol or fructose and xylitol-fructose mixtures, but after sorbitol it remained constant. A slight insulin release only after xylitol. Cholesterol decreased after sorbitol, fructose and xylitol. Neutral fats increased slightly after fructose, but not after sorbitol or xylitol. Aspartate aminotransferase, lactate dehydrogenase and alkaline phosphatase were unchanged. Fructose increased lactate more than sorbitol or xylitol. Water balance was positive after fructose and glucose, but positive or negative after sorbitol or xylitol, depending on dose. Utilization at a rate of 0.25 g/(kg × hr): fructose, 99.3%; sorbitol, 88.6%; xylitol, 84.4%.
*Donahoe and Powers (1974) [649]	Man. Intravenous infusion (the doses of xylitol ranged from 0.59 to 1.43 g/kg for one day, or from 0.56 to 3.13 g/kg in multiple infusion for 3 days)	The data suggested a positive dose response between xylitol and serum uric acid. Single 2-h infusions of 25 or 100 g xylitol increased serum uric acid, but the levels were within the normal range. Two infusions per day caused an elevation above the normal range. The 24-h urinary excretion of uric acid was doubled after two 100-g infusions per day. Bilirubin conjugation was interfered and total bilirubin levels were increased. The authors provided two explanations to the discrepancy between their own and other findings: (1) correct timing and sufficient frequency of blood sampling; (2) different dosing schedule and differences in individual sensitivity (see chapter 7.31).
1975		
*Mehnert, Dietze and Haslbeck (1975) [527]	Man, healthy, intravenously, 0.5 g/(kg × hr) for 4 hours.	Only slight changes in blood glucose and insulin after fructose, sorbitol or xylitol. Lactate was increased after sorbitol and fructose, whereas glucose and xylitol produced only slight changes.
Woods (1975) [376]	Rat, liver perfusion.	Xylitol loading may cause a large depletion of ATP, total adenine nucleotides and P_i, accompanied by the accumulation of a phosphorylated intermediate. The fall in hepatic P_i and the activation of AMP-degrading, largely irreversible enzymes, may explain the depletion in adenine nucleotides. The author emphasized that infusion of high amounts of xylitol is not without risk (see chapter 7.3).
*Heaf and Galton (1975) [338]	Investigation of non-diabetic and diabetic patients without administration of xylitol.	The daily urinary excretions of glucose, sorbitol and inositol were raised in diabetics. In lens of diabetics there was an increase in glucose concentration, whereas fructose, sorbitol or inositol were not increased. Urinary pentitol excretion was not affected by the presence of diabetes.

Authors	Description of study	Findings
*Heuckenkamp and Zöllner (1975) [528]	Man, intravenously, 0.5 g/(kg×hr).	Considerable loss of xylitol in urine (amounting to 22% of the dose), because pentitols are not reabsorbed within the kidney tubules after filtration of the glomerular membrane. This was considered to be a general feature of polyols. 0.3–0.5 g/(kg×hr) infusion increased serum urate concentration. Fructose and glucose were equally well utilized in metabolically healthy subjects.
*Matzkies, Heid, Fekl and Berg (1975) [529]	Man, infusion of a mixture of fructose, glucose and xylitol (2:1:1) at a rate of 1.5 g/(kg×hr) as a 20% solution free of electrolytes (over 1 hr).	Serum glucose was increased from 80 to 130 mg/100 ml, insulin from 17 to 78 μU/ml. The rapid metabolism of carbohydrates was responsible for the significant lactate and pyruvate increase. No change in Cl and K, Na, Ca and P were decreased at the end of the infusion.
*Berg, Matzkies, Heid, Fekl and Conolly (1975) [530]	Man, infusion of a mixture of fructose, glucose and xylitol (2:1:1) at a rate of 0.5 g/(kg×hr), corresponding to 0.125 g/(kg×hr) of xylitol (over 12 hrs).	No clinical side effects. Changes in all metabolites were within the physiological range. Utilization of glucose: 100% (98.8% for fructose and 87.9% for xylitol). Electrolyte loss amounted to 2.36 mval K/hr and 10.15 mval Na/hr. At a rate of 0.5 g/(kg×hr) a concentration of 14.3 mval K/l and 61.5 mval Na/l is required.
*Berg, Matzkies, Heid and Fekl (1975) [369]	Man, infusion with an electrolyte solution containing 24% carbohydrate (fructose, glucose and xylitol; 2:1:1). Rate: 0.6 g/(kg×hr).	Slight increase of glucose and insulin. No increase in serum oxalate. Infusion of an amino acid mixture containing glycine: oxalate excretion was decreased during infusion. Lactate was not changed, but β-hydroxybutyrate decreased by the end of the infusion (= antiketogenic effect). Only slight modifications in electrolyte concentrations. Transaminases and bilirubin were not changed. Utilization of sugars: glucose, 100%; fructose, 99%; xylitol, 88%.
*Mäkinen and Scheinin (1975) [302]	Man, oral feeding of ca. fifty subjects for 2 years using a very versatile assortment of foodstuffs (included one diabetic subject). Blood chemistry was monitored seven times during the period (the last analysis at the 22-month phase). Mean monthly consumption of xylitol 1.5 kg. The highest daily doses 200–400 g.	No clinically significant changes in the following serum or blood parameters: Ca, Mg, Na, K, P_i, bilirubin, ascorbate, hemoglobin, leukocytes, sedimentation rate of erythrocytes, alkaline phosphatase, both transaminases, lactate dehydrogenase, amino acids, isoelectric focusing of serum proteins. Serum amylase decreased, but the values were still at normal level. Transient diarrhea in half of the subjects. Adaptation to even high doses developed within 1–3 weeks. Values of pregnant and the diabetic subjects were normal. These studies did not show any difference between the consumption of moderate quantities of sucrose, fructose or xylitol.

Authors	Description of study	Findings
*Mäkinen, Mielityinen and Scheinin (1975) [303]	Disc electrophoresis of some serum proteins in the above study.	No significant differences between the test groups were found.
*Huttunen, Mäkinen and Scheinin (1975, 1976) [294, 531]	Additional serum analyses in the 2-year Turku study, including assay of urine uric acid.	No clinically significant differences between the test groups in the following parameters: serum insulin, lactate, pyruvate, cholesterol, triglycerides, glucose and urate, and urine urate.
*Matzkies (1975) [370]	Man, long-term infusion (6–12 hrs), up to 0.75 g/(kg × hr) of glucose (0.50 for sorbitol; 0.25 for fructose or xylitol).	Up to the dosages indicated a steady state of substrates and metabolites was found. Exceeding of dosage limits: glucose, hyperglucosemia; sorbitol, rise in sorbitol blood level; fructose, rise in lactate; xylitol, clinical side effects. Simultaneous application of glucose, fructose and xylitol: utilization of glucose rises, wheras the utilization of fructose and xylitol remains on levels as if these sugars were infused alone.
Wang, Oshinsky, Lantin and van Eys (1975) [552]	Rabbit, infusion of a 25% solution of xylitol at a rate of 2 g/kg in normal and oxythiamine-treated fasted rabbits.	A transient but modest increase in plasma oxalate was observed in both groups. The urinary oxalate levels remained normal. There was no significant difference of plasma or urinary oxalate levels between the groups. More severe liver toxicity was found in xylitol-oxythiamine-treated animals and the retention time of plasma xylitol was longer. Glucose, at an equivalent dose, resulted in higher plasma oxalate than xylitol. No experimental evidence for a relation between oxalate accumulation and xylitol toxicity was found.
Thomas, Edwards and Edwards (1975) [532]	Dog, intravenously, 1 g/kg for 4.5–5.5 hours. Isolated rat liver cells. Rat, intravenously, 1 g/(kg × hr) for 60 hours, preceded by an infusion with a 50% solution for 48 hours.	Glucose, fructose, sorbitol and xylitol increased plasma osmolality in dog, most with sorbitol and xylitol, indicating that the two polyols are not as effectively taken up by the intracellular compartment as glucose or fructose. All sugars caused acidosis in dogs. Xylitol and sorbitol altered plasma lactate/pyruvate ratio. Rat isolated liver cells behaved in an opposite way. Sorbitol and xylitol increased the level of α-glycerophosphate in dog. B_1- and B_6-deficient rats: oxalate excretion increased.

Authors	Description of study	Findings
1976		
Oshinsky, Wang and van Eys (1976) [268]	Rabbit, infusion in normal and 4-deoxypyridoxine-treated animals, at 2 g/kg as a 25% solution.	A small increase in plasma oxalate from 20 to 60μg/ml was observed in 30 minutes. Urinary oxalate levels were normal. No significant difference in plasma or urinary oxalate levels between the groups. Plasma xylitol retention was similar in both groups. Glucose, at an equivalent dose in 4-deoxy-pyridoxine-treated rabbits, resulted in higher plasma oxalate levels than xylitol. Oral xylitol (5 g/kg) gave no urinary oxalate rise in treated animals. The findings suggested no relationship between oxalate accumulation and xylitol toxicity.
*Dubach (1976) [555]	Man (diabetic, 43–80 years, of both sexes). (A) Intravenously 0.5 g/kg in 10 minutes. (B) Infusion; 0.5 g/(kg × hr) in 2 hours.	(A) Sorbitol caused a higher peak and a larger total insulin secretion than xylitol. (B) Insulin secretion was more pronounced with xylitol than with sorbitol. Intravenously applied xylitol and sorbitol increased blood glucose. The xylitol-caused increase in blood glucose could be avoided by pretreatment with tolbutamide, although the secretion of insulin remained unchanged. In B: Uric acid was significantly increased by xylitol but not by sorbitol. The difference between sorbitol and xylitol was almost significant ($p<0.05$). Both xylitol and sorbitol increased bilirubin ($p<0.05$; no differences between the two polyols).
*Matzkies and Berg (1976) [566]	Man, six healthy adults. Intravenously 0.5 g/(kg × hr) for 12 hours, or eight adults 0.6 g/(kg × hr) + amino acids for 12 hours.	The mixture of fructose, glucose and xylitol (2:1:1) could be infused at a rate of 0.5 g/(kg × hr) for 12 hours without any greater side effects.
*Bickel (1976) [557]	(A) Man. Infusion in thirty-two patients after gastrointestinal surgery 0.5 g/(kg × hr) over 6 hours. Mixtures of glucose, fructose and xylitol, or of fructose, glucose and xylitol (2:1:1). (B) Ninety-two patients at a rate of 0.5 g/(kg × hr). (C) Eight patients at a rate of 0.5 g/(kg × hr) for 6 hours. Either glucose, fructose, xylitol, or their mixture.	(A) The undesired effects of glucose, fructose and xylitol (when infused in the immediate postoperative period at rates necessary for parenteral nutrition) can be reduced significantly if the substances are infused as a mixture. The mixture forms a low osmotic load in the extracellular space. (B) Base excess was decreased only moderately by glucose or the mixture, whereas fructose or xylitol showed a greater effect. This is important, as 75% of the mixture was made up of fructose and xylitol. (C) P was reduced during the infusions of all carbohydrates, but the greatest decrease was obtained with xylitol. This was also found in normal subjects [472]. Marked increase in serum urate was found only in some patients of the xylitol group; in some patients the values were even reduced. These changes were in accordance with those of Dölp et al. [593]. It was concluded that the xylitol-induced increase of serum urate in the postoperative phase is not more distinct than in normal subjects (as also presented by Förster [352]).

Authors	Description of study	Findings
*Willgerodt and Beyreiss (1976) [558]	Man (seventy-eight neonates and premature infants, aged 24 hrs to 8 days). Continuous intravenous infusion of xylitol or sorbitol. 0.5–1.0 g/(kg × hr) over 2–3 hours.	Blood glucose was clearly increased. At a rate of 0.5 g a steady state was established for both polyols after 60–90 minutes. A maximal intake rate of 0.25 g/(kg × hr) should not be surpassed for long-term intravenous nutrition of neonates. A rapid (2–3 min) infusion of xylitol also increased blood glucose, whereas sorbitol led to a delay of the increase of glucose. Rapid infusion of both polyols at a rate of 1.0 g/(kg × hr) increased blood lactate significantly. It was suggested that rapid infusion must be avoided, but polyols can in part be used instead of glucose in parenteral nutrition in the neonatal period.
Thomas, Hannett, Chalmers, Rofe, Edwards and Edwards (1976) [355]	Rat, normal or deficient in vitamins B_1 and B_6. Intravenous infusion up to 5 days (0.75 ml/hr). The concentration was slowly increased from 5 to 40% (w/v). Final dose: 1.2 g/(kg × hr).	The urinary excretion of oxalate, glyoxylate and glycine increased considerably in rats deficient in vitamin B_6, when the rats were infused with xylitol. Correction of the vitamin deficiency resulted in a return to normal excretion. It was suggested that xylitol metabolism produces an increased activity in pathways leading to oxalate formation in vitamin B_6-deficient rats.
*Paulini (1976) [559]	Man. Forty-five deceased persons after infusion of carbohydrates.	No intrarenal crystal deposition in thirty-three cases. In twelve cases depositions of varying intensity were found. The latter had received considerable quantities of xylitol within 10 days (1 kg to 3.2 kg). The author stated that there may be a possible connection between infusion of xylitol-containing solutions and intrarenal crystal deposition.
*Pesch, Krampf, Menzel, Weiland, Eidam, Prestele and Heid (1976) [560]	Man (and rabbit). Histochemical and biochemical studies on three hundred patients who had died in the recovery room after infusion of carbohydrates.	Independent of any infusions, deposits of calcium oxalate crystals were found in the presence of kidney-specific diseases. Xylitol infusions of 0.4 g/kg, or, in individual cases, of not more than 500 g in 7 days, had no influence on the appearance of calcium oxalate deposits.
*Watts, Hauschildt, Chalmers and Lawson (1976) [561]	Man. Ten persons undergoing gastric or biliary tract surgery. Five were infused with glucose, five with xylitol at a rate of 0.25 g/(kg × hr) with a 10% solution (w/v) over a period of 120 hours (no sugar infused on operation day).	Four of the five xylitol-infused patients had hyperglycollic aciduria. Three of the glucose-infused ones had hyperlactic aciduria. No hyperoxaluria was found. Four of the xylitol-infused patients excreted more tetronic acids than the other group. Threonic acids prevailed in the xylitol group, erythronic acids in the glucose group. It was concluded that the transketolase pathway may be overloaded during xylitol infusion.

Authors	Description of study	Findings
*Dölp and Paulini (1976) [562]	Man. Twenty-two persons divided into two groups: 7 g sorbitol or xylitol/(kg×day). Infusion during 3–10 days.	The excretion of oxalate was significantly higher in the xylitol group as compared to the sorbitol group. One died patient of the xylitol group had massive oxalate deposits in the brain and kidney and 4-fold higher glycine level in plasma as compared to normal values. No oxalose was, however, later detected after the use of combinations of carbohydrates supplemented with amino acids under control of the acid-base balance (see chapter 7.3). See also [593].
*Ahnefeld, Halmágyi and Milewski (1976) [563]	Man. (A) Sixty patients in the postoperative phase. Infusion of sugars or their mixtures, 2–3 g/(kg×day). (B) Fifty patients (at labour) at a rate of 0.3 g/(kg×hr) at least 3 hours (sugars or their mixtures). (C) Ten intensive-care patients, 0.5–1.0 g/(kg×hr).	Monotherapy with any of the carbohydrates was less sufficient than with a combination of these substrates.
Froesch (1976) [564]	Rat. Injection of 1 μCi of universally labelled xylitol and other sugars.	10 minutes after the injection of labelled xylitol or sorbitol more than 80% of the ^{14}C of plasma was in the form of ^{14}C-glucose. This percentage increased to nearly 100% at later intervals. In man the conversion of intravenously administered labelled sorbitol and xylitol to plasma glucose was similar but slower than in the rat. The author claimed that a large portion of the sugar substitutes is rapidly converted to glucose and metabolized further only with the help of adequate amounts of insulin and that insulin-independent utilization is not met by xylitol or sorbitol. Sorbitol and xylitol are, according to the author, experimental tools and they do not have any place in parenteral nutrition (see chapter 7.3).
*Göschke, Leutenegger and Allgöwer (1976) [565]	Man. Twenty-four cholecystectomized or vagotomized patients. Infusion of glucose or a combination of glucose, fructose and xylitol (1:2:1) for 5 days at a rate of 1.42–7.14 g/(kg×day).	Tolerance was good in both groups, but urinary losses of infused carbohydrates were higher in the group infused with the mixture. It was concluded that after surgery of intermediate magnitude the combination of glucose, fructose and xylitol does not offer any advantage over glucose alone. However, in severely ill patients with pronounced glucose intolerance, studies with sugar substitutes appear of interest.

Authors	Description of study	Findings
*Brand and Quadflieg (1976) [567]	In vitro studies with human erythrocytes.	Xylitol was utilized by erythrocytes at a rate of 0.6 µmoles xylitol/ml packed cells in 60 minutes at physiological pH. The rate was lower at pH 7.8. Utilization of xylitol was coupled with the breakdown of 2,4-diphosphoglycerate to lactate in order to generate NAD. Human erythrocytes metabolized xylitol via the pentose phosphate and the glycolytic pathway. All atoms taken up from xylitol were recovered in products like lactate and CO_2.
*Matzkies and Berg (1976) [556]	4–8 healthy persons. Intravenously a mixture of fructose, glucose and xylitol (2:1:1) at a rate of 0.5 g/(kg × hr) for 6 hours, or at a rate of 0.6 g/(kg × hr) for 12 hours with a concomitant application of amino acids.	Total clearance (in ml/(kg × min): Fructose-glucose-xylitol: 42 Fructose-glucose-xylitol + amino acids: 104 Glucose alone [given at 0.125 g/(kg × hr)]: 14.7
Förster (1976) [580]	Rat. Liver perfusion tests. An assessment of previous studies [285, 292].	In a typical perfusion test the concentration of glucose remains constant in the perfusion solution even at high glucose concentration, indicating that glucose may not be metabolized by the isolated rat liver. Sorbitol, xylitol and fructose are taken up quickly. One of the main products is glucose (60–70% of xylitol is converted, 45–50% of sorbitol or fructose) which is required especially by the brain and the blood cells. Xylitol can reduce nitrogen catabolism by the liver.
Mader and Reinauer (1976) [553]	Rat. Perfusion of the heart muscle. The degradation of labelled sugars to CO_2 was measured.	Only 3–4% of the oxygen consumption was referred to xylitol utilization. It was concluded that with xylitol as the only substrate the heart is working under substrate deficiency conditions and that diabetes mellitus does not improve the xylitol utilization in the rat heart muscle.
Poutiainen, Tuori and Sirviö (1976) [576]	Cow. Investigations of the activity of rumen microbes to ferment pure polyols or a mixture of various polyols.	After 8 hours' incubation over 80% of the polyols (xylitol, arabinitol, mannitol, galactitol, sorbitol, rhamnitol) remained unfermented. After 24 hours less than 20% of the xylitol and arabinitol had been fermented, but of the galactitol and mixed polyols 60% and almost all of the mannitol and sorbitol had been fermented. The polyols not fermented in the rumen were nervertheless absorbed and metabolized by the cow.

Authors	Description of study	Findings
*Sestoft and Gammeltoft (1976) [610]	Man. Liver biopsies of patients who underwent cholecystectomy were performed after xylitol infusion (25 or 50 g at a constant rate during 30 min).	50-g dose decreased the hepatic ATP concentration from 2.75 to 0.25 μmole/g liver and the concentration of P_i from 3.6 to 1 μmole/g. The hepatic content of adenine nucleotides was reduced. L-Glycero-3-phosphate increased, but glucose, lactic acid and ketone bodies remained unchanged (see chapter 7.3).
*Seino, Taminato, Inoue, Goto, Ikeda and Imura (1976) [594]	Man. Intravenous infusion of xylitol. Priming dose: 0.25 g/kg (50% solution) in 2 minutes. This was followed by infusion of 10% xylitol at 15 mg/(kg × min) over 90 minutes. 30 g of L-arginine-HCl (in 300 ml water) was infused over a period of 45 minutes starting 45 minutes after xylitol.	Xylitol was suggested to have an inhibitory effect on both basal and arginine-stimulated glucagon secretion, while it enhances insulin secretion. These effects can be possible through metabolic pathways in both α- and β-cells. Xylitol may also interact with membrane receptors of α- and β-cells which influence the arginine-associated glucagon and insulin secretion.
Demetrakopoulos and Amos (1976) [595]	Chick embryo fibroblast cultures. Uptake of sugars was studied using radiolabelled compounds.	Xylitol and D-xylose supported excellent growth from small inocula of chick embryo fibroblasts, NIL Syrian hamster, Chinese hamster ovary cells and human skin fibroblasts. Xylitol, D-xylose and D-ribose enter the cells by simple diffusion and do not exhibit any competition with other pentoses or hexoses.
*Hauschildt, Chalmers, Lawson, Schultis and Watts (1976) [650]	Man. Parenteral nutrition at the postoperative phase (0.25 g/kg × h). During a 53-h period two 5-h infusions were performed; Ringer's solution was infused before, between and after the xylitol infusions. Glucose at the same rate as control. In another experiment xylitol or glucose were infused in a 120-h study between 0 and 24 h, 24 and 48 h, 72 and 96 h and 96 and 120 h (at the above rate).	The thiamine status of the subjects was normal. Some subjects showed slight pyridoxine deficiency, but there were no untoward effects of xylitol infusion. Blood and urine oxalate remained within the normal range. Xylitol and glucose decreased the plasma and urinary orthophosphate and urinary pyrophosphate. Plasma pyrophosphate and Ca and urinary Ca were unaltered. Xylitol infusion was associated with an abnormal glycolic aciduria and tetronic aciduria, but no increase in oxalate excretion was observed. It was suggested that patients with renal failure infused rapidly with large doses of xylitol are unable to excrete all glycolic acid formed with a subsequent accumulation of oxalate. Wang, Oshinsky and van Eys [652] subsequently discussed the oxalate levels reported by Hauschildt et al. [650]. The normal values in the laboratory of Wang et al. were 397 ± 287 μg/100 ml plasma (mean ± 1 SD, n = 20), which are higher than those reported by the previous [650] authors. Animal studies of Oshinsky, Wang & van Eys [651] indicated the unlikelihood that oxalate formation was the cause of toxicity as asserted by the Australians [265].

Authors	Description of study	Findings
Mäkinen, Bowen, Dalgard and Fitzgerald (1976, 1977) [206, 244]	Monkey *(Macaca mulatta)*, oral feeding (30 g/day) for 3 days.	Consumption of a xylitol diet increased significantly the activity of lactoperoxidase in parotid and submandibular saliva when compared to sucrose consumption. Secretion of total proteins was also increased. The concentration of phosphate was not affected in parotid saliva, but xylitol consumption slightly decreased the concentration in the submandibular saliva when compared to sucrose intake.
1977		
Mäkinen, Bowen, Dalgard and Fitzgerald (1977) [244]	Monkey *(Macaca mulatta)*, gastric intubation of 2.5 g xylitol or sorbitol daily for 3 days.	Intubation of the 2.5-g dose in three separate and equal portions increased slightly parotid lactoperoxidase activity compared to intubation as a single dose/day. No significant differences between sorbitol and xylitol were observed with regard to parotid saliva lactoperoxidase activity.
Bird, Baum, Mäkinen, Bowen and Longton (1977) [245]	Monkey *(Macaca mulatta)*, oral feeding (as in [244]).	The activity of parotid saliva amylase was significantly increased on xylitol diet when compared to sucrose consumption. The concentration of total proteins was also increased.
*Ylikahri and Leino (1977) [554]	Man. Seven healthy volunteers. Administration of xylitol per os (1.0 g/kg) with or without intravenous ethanol (0.8 g/kg) during the first hour of experiment. Glucose as the control.	Xylitol did not affect the rate of ethanol elimination, but during the infusion of ethanol the concentrations of blood xylitol were 5–10 times higher than after xylitol only. It was concluded that ethanol inhibits the elimination of xylitol. Xylitol did not modify significantly the concentration of glucose in blood, nor the ethanol-induced changes in blood lactate and pyruvate.
Bjondahl and Virkki (1977) [589]	Mouse. Gastric intubation of low quantities of xylitol (or a polyol mixture) together with tritiated vitamin A_1 (a mixture of 'cold' vitamin A, D and E was simultaneously given).	Administration of xylitol enhanced the transfer of the label from the gastrointestinal tract to serum and liver. The possibility exists that the absorption of vitamin A was stimulated.
*Förster (1977) [602]	Man. Oral administration.	Confirmation of previous findings that oral administration of xylitol (at total daily dosage = 100 g in adults) lacks adverse reactions.

Authors	Description of study	Findings
*Mäkinen, Mäkinen, Söderling and Tenovuo (1977) [603]	Man. Oral administration (single dose, 0.6 g/kg). Isolation of granulocytes, lymphocytes and erythrocytes at 0, 1, 2 (or 3) hours after administration.	Xylitol did not have any detectable effect on the activity of leukocyte peroxidase, lactate dehydrogenase, alcohol dehydrogenase, glucose 6-phosphate dehydrogenase, or on the concentration of ionized iodine. Erythrocyte catalase and glucose 6-phosphate dehydrogenase activity did not differ from controls either. The same concerned whole saliva peroxidase and the concentration of ionized iodine and SCN^- ions. Serum lactate dehydrogenase, lactic acid and pyruvic acid did not differ from controls.
Mäkinen, Tuori, Poutiainen and Hämäläinen (1977) [604]	Cow. Oral feeding with a polyol mixture (containing chiefly xylitol, sorbitol, mannitol and arabinitol: 0.5 kg/animal daily over a period of 11 weeks).	No clinically significant changes compared to controls in the following parameters: - Serum transaminases, α-amylase, phosphatases, bilirubin, glucose, protein, P_i, Fe, Ca, Mg, K, Na, sulphate, sialic acid, cholesterol, amino acids. - Milk alkaline phosphatase, protein, glucose, Fe, Ca, Na, Mg, K. - Lacrimal fluid protein, lactoperoxidase, aminopeptidase. - Whole saliva protein, peroxidase, amylase. The consumption of the polyol mixture maintained a higher and more constant lactoperoxidase activity in milk than the control diets.
*Förster, Boecker and Walther (1977) [618]	Man. Diabetic children, aged 5–15 years. Oral administration over a period of 4 weeks (30 g/day).	The subjects tolerated xylitol fairly well and only one child terminated the trial before the end due to diarrhea. A significant increase of serum uric acid concentration measuring 1 mg/100 ml was the main side effect which was favoured by the fact that these subjects did not use sucrose (a sucrose-free period is normally required to find xylitol-associated hyperuricemia). In metabolically healthy children sucrose-induced hyperuricemia is also to be expected. Thus this effect does not obviously have any pathophysiological significance. Xylitol is suited in diabetic diet if the limited dose is observed.
Brin, Gabriel and Machlin (1977) [619]	Rat.The animals were fed a semi-purified thiamine-deficient diet with 8% sorbitol or 8% xylitol for 2 weeks, after which the polyols were increased to 2% for 2 more weeks.	Sorbitol produced diarrhea, while xylitol animals remained normal. At high doses sorbitol was suggested to produce a malabsorption syndrome. Xylitol spared thiamine more effectively than sorbitol, presumably by increasing intestinal synthesis and absorption of the vitamin.

Authors	Description of study	Findings
*Hoerman (1977) [620]	Man. Oral administration of 14 g xylitol daily for a period of 8 weeks (in the form of chewing gum).	No untoward gastrointestinal effects were noted and the oral tissues remained normal in all subjects.
Asano, Greenberg, Wittmers and Goetz (1977) [621]	Dog. Healthy female animals, fasted. Intravenous infusion of xylitol or glucose for 50 minutes (25–50 g/100 ml) Total dose of sugar/kg: 0.7–4.0.	The levels of plasma insulin during xylitol infusion were as high or higher than those during glucose infusion, with increases in arterial levels of xylitol, which were equal to or less than those in glucose. More xylitol was lost in the urine, but the overall uptake was $\geqq 70\%$. The ability of xylitol to stimulate both phases of insulin release and its rapid uptake by tissues may result from its homology with α-D-glucopyranose. The authors suggested that there is a glucoreceptor on the surface of the β-cell, with stereospecifity for α-D-glucopyranose or other similar molecules.
Wang, Oshinsky, Lantin, Ukab and van Eys (1977) [654]	Rabbit. Infusion with either xylitol or glucose [40 mg/(kg × min)] of fasted normal, oxythiamine-treated and 4-deoxypyridoxine-treated animals.	Oxalate production from xylitol or glucose was on a similar order of magnitude, and negligible. Xylitol induced slight alterations in P_i, α-glycerophosphate, lactic acid and allantoin levels, but the fluctuations were infusion rate-dependent. The results suggested that any toxicity associated with xylitol administration is not a consequence of oxalate production. The safety margin of xylitol in rabbits was considered wide.
Hannett, Thomas, Chalmers, Rofe, Edwards and Edwards (1977) [655]	Rat (normal and animals deficient in thiamine or vitamin B_6). Infusion over a period of 5 days at a rate of 1 g/(kg × h).	Urinary excretions of oxalate, glyoxylate and glycine increased significantly in vitamin B_6-deficient rats infused with xylitol when compared with other groups (glucose, fructose, sorbitol). Oxalate was shown to be derived directly from xylitol in vitamin B_6-deficient rats. The authors suggest that vitamin B_6-deficiency may have been a factor which contributed to oxalate crystal deposition in some patients infused with xylitol.
Oshinsky, Wang and van Eys (1977) [651].	Rabbit. Infusion with either xylitol or glucose [2 mg/kg, corresponding to 40 mg/(kg × min)] of fasted normal, oxythiamine-treated and 4-deoxypyridoxine-treated animals.	Transient thiamine- or pyridoxine-deficient states were observed in the antivitamin-treated animals. During the first 24 hrs following infusion with either carbohydrate, urinary oxalate remained normal. The results suggested that oxalate production from xylitol is negligible. Consequently, any toxicity related to xylitol administration is not a consequence of oxalate production.

Authors	Description of study	Findings
*Leutenegger, Göschke, Stutz, Mannhart, Werdenberg, Wolf and Allgöwer (1977) [653]	Man, total parenteral nutrition, postoperatively with a combination of glucose, fructose and xylitol (4 days, 600 g of the mixture per 24 h; 600 g glucose/24 h as control). The proportions of the GFX mixture were 1:2:1.	The mean plasma glucose was significantly ($p < 0.005$) decreased (154.2 ± 19.5 mg/100 ml) as compared to infusion with glucose alone (193.9 ± 15.0 mg/100 ml). The required dosage of exogenous insulin was significantly lower during infusion of the mixture. The renal carbohydrate losses were 0.85% during infusion with glucose and 1.7% during infusion of the mixture. No clinical side effects during infusion on either of the following parameters: pH, base excess, lactate, pyruvate, free fatty acids, insulin, Na, K, Cl, Mg, P_i, bilirubin, alkaline phosphatase, serum transaminases. It was concluded that administration of the sugar mixture is justified when hyperglycemia during infusion of glucose is difficult to control without insulin.

11. References

1 E. Fischer and R. Stahel: Ber. Dtsch. Chem. Ges. *24,* 528 (1891).
2 M.G. Bertrand: Bull. Soc. Chim. Paris, *5,* 554 (1891).
3 M.L. Wolfrom and E.J. Kohn: J. Am. Chem. Soc. *64,* 1739 (1942).
4 J.F. Carson, S.W. Waisbrot and F.T. Jones: J. Am. Chem. Soc. *65,* 1777 (1943).
5 J.F. Carson and W.D. Maclay: J. Am. Chem. Soc. *67,* 1808 (1945).
6 R.C. Hocket and C.S. Hudson: J. Am. Chem. Soc. *57,* 1735 (1935).
7 G.F. Hewitt and C.S. Hudson: J. Am. Chem. Soc. *69,* 921 (1947).
8 A. Gilman, F.S. Philips, R.P. Allen and E.S. Koelle: J. Pharmacol. Exp. Ther. Suppl. *87,* 85 (1946).
9 A.H. Lawton, F.J. Brady, AT. Ness and W.T. Haskins: Am. J. Trop. Med. Hyg. *25,* 263 (1945).
10 W. Traube and F. Kuhbier: Ber. Dtsch. Chem. Ges. *69,* 2661 (1936).
11 W. Traube and F. Kuhbier: German Patent 673,874, March 30 (1939).
12 R.S. Tipson and L.H. Cretcher: J. Org. Chem. *8,* 95 (1943).
13 Beilsteins Handbuch der Organischen Chemie, 4. Aufl. E III, Vol. 1/II. Syst. Nr. 54, p. 2379. Springer, Berlin 1958.
14 U. Manz, E. Vanninen and F. Voirol: Food R.A. Symposium on Sugar and Sugar Replacements. London, October 1973.
15 F. Kracher: Kakao + Zucker *27,* 68, 108 (1975).
16 R.L. Lohmar, in: The Carbohydrates, Chemistry, Biochemistry, Physiology, p. 241. Ed. W. Pigman. Academic Press, New York 1962.
17 J. Böseken: Advanc. Carbohydr. Chem. *4,* 189 (1949).
18 K. Kratzl: Monatsh. Chem. *94,* 106 (1963).
19 C. Toniolo, G.M. Bonora and A. Fontana: Int. J. Pept. Protein Res. *6,* 283 (1974).
20 O.P. Chilson, L.A. Costello and N.O. Kaplan: Fed. Proc. *24,* Suppl. 15, 55 (1965).
21 F. Rappaport, I. Reifer and H. Weinmann: Mikrochim. Acta *1,* 290 (1937).
22 C.D. West and S. Rapoport: Proc. Soc. Exp. Biol. Med. *70,* 141 (1949).
23 J. Washüttl, P. Riederer, E. Bancher, E. Wurst and K. Steiner: Z. Lebensm. Unters. Forsch. *155,* 77 (1974).
24 F.D. Gauchel, G. Wagner and K.H. Bässler: Z. Klin. Chem. Klin. Biochem. *9,* 25 (1971).
25 A. Yoda: J. Chem. Soc. Jpn. *73,* 18 (1952).
26 P. Godin: Nature *174,* 134 (1954).
27 J. Washüttl, P. Riederer and E. Bancher: J. Food Sci. *38,* 1262 (1973).
28 K. Kratzl and H. Silbernagel: Naturwissenschaften *50,* 154 (1963).
29 C. Maurer and P. Christophis: Z. Klin. Chem. Klin. Biochem. *11,* 535 (1973).
30 K.H. Bässler, V. Unbehaun and W. Prellwitz: Biochem. Z. *336,* 35 (1962).
31 A. Przybilka and B. Linke: Pharm. Ztg. *116,* 1061 (1971).
32 K.K. Mäkinen, to be published.
33 C. Aminoff, in: Sugars in Nutrition, p. 135. Eds. H.L. Sipple and K.W. McNutt. Academic Press, New York 1974.
34 W.J. Nickerson and R.G. Brown: Adv. Appl. Microbiol. *7,* 225 (1965).
35 H. Onishi: Bull. Agr. Chem. Soc. Jpn. *24,* 131 (1960).
36 H. Onishi and M.B. Perry: Can. J. Microbiol. *11,* 929 (1965).

37 H. Onishi and T. Suzuki: Agr. Biol. Chem. *30,* 1139 (1966).
38 H. Onishi and T. Suzuki: J. Bacteriol. *95,* 1745 (1968).
39 H. Onishi and T. Suzuki: Appl. Microbiol. *16,* 1847 (1968).
40 H. Onishi, N. Saito and I. Koshiyama: Agr. Biol. Chem. *25,* 124 (1961).
41 S.M. Partridge: Nature *164,* 443 (1949).
42 R. Prince and T. Reichstein: Helv. Chim. Acta *20,* 101 (1937).
43 S. Roseman, R.H. Abeles and A. Dorfman: Arch. Biochem. Biophys. *36,* 232 (1952).
44 J.F.T. Spencer: Prog. Ind. Microbiol. *7,* 1 (1968).
45 J.F.T. Spencer and H.R. Sallans: Can. J. Microbiol. *2,* 72 (1956).
46 J.F.T. Spencer, J.M. Roxburgh and H.R. Sallans: J. Agric. Food Chem. *5,* 64 (1957).
47 H. Onishi and T. Suzuki: Appl. Microbiol. *18,* 1031 (1969).
48 C. Chiang, C.J. Sih and S.G. Knight: Biochem. Biophys. Acta *29,* 664 (1958).
49 T. Suzuki and H. Onishi: Agr. Biol. Chem. *31,* 1233 (1967).
50 V.I. Scharkow: Chem.-Ing.-Tech. *35,* 494 (1963).
51 M. Frèrejaque: C. R. Acad. Sci. *217,* 251 (1943).
52 D.L. Corina and K.A. Munday: J. Gen. Microbiol. *69,* 221 (1971).
53 I. M. Morrison: Biochem. Soc. Transact. *3,* 992 (1975).
54 W.H. Lee: Appl. Microbiol. *15,* 1206 (1967).
55 G.W. Strandberg: J. Bacteriol. *97,* 1305 (1969).
56 D. Rast: Ber. Schweiz. Bot. Ges. *76,* 176 (1966).
57 K. Horikoshi, S. Iida and Y. Ikeda: J. Bacteriol. *89,* 326 (1965).
58 K.B. Helle and L. Klungsoeyr: Biochim. Biophys. Acta *65,* 461 (1962).
59 J.H. Birkinshaw, J.H.V. Charles, A.C. Hetherington and H. Raistrick: Philos. Trans. R. Soc. London *B220,* 153 (1931).
60 F. Gehring: Dtsch. Zahnärztl. Z. *29,* 769 (1974).
61 K.K. Mäkinen: Int. Dent. J. *22,* 363 (1972).
62 K.K. Mäkinen, in: Sugars in Nutrition, p.645. Eds. H.L. Sipple and K.W. McNutt. Academic Press, New York 1974.
63 A.J.W. McKendrick, in: Dental Plaque, p.297. Ed. W.D. McHugh. D.C. Thomson, Dundee 1970.
64 H.V. Jordan, in: Streptococcus mutans and Dental Caries, p.45. Department of Health, Education and Welfare Publ. No. (NIH) 74-286, Bethesda 1973.
65 R.G. Schamschula, P.H. Keyes and R.W. Hornabrook: J. Am. Dent. Assoc. *85,* 603 (1972).
66 I.L. Shklair and H.J. Keene: 51st Session Int. Assoc. Dental Res., Abstract No.594 (1973).
67 I.L. Shklair, in: Streptococcus mutans and Dental Caries, p.7. Department of Health, Education and Welfare Publ. No. (NIH) 74-286, Bethesda 1973.
68 J.D. de Stoppelaar, V. van Houte and O. Backer Dirks: Caries Res. *3,* 190 (1969).
69 G.G. Kozlowski, I.L. Shklair, H.J. Keene and J.A. Levine: 51st Session Int. Assoc. Dental Res., Abstract No.552 (1973).
70 B. Krasse, H.V. Jordan, S.Edwardsson, I.Svensson and L. Trell: Arch. Oral Biol. *13,* 911 (1968).
71 R.J. Gibbons and W.J. Loesche: Arch. Oral Biol. *12,* 1013 (1967).
72 H.V. Jordan, H.R. Englander and S. Lim: J. Am. Dent. Assoc. *78,* 1331 (1969).
73 H.M. Stiles, C.J. Donnelly, L.A. Thomson and J.A. Brunelle: 50th Session Int. Assoc. Dental Res., Abstract No.681 (1972).
74 N.W. Littleton, S. Kakehashi and R.J. Fitzgerald: Arch. Oral Biol. *15,* 461 (1970).
75 I.L. Shklair, H.J. Keene and L.G. Simonson: J. Dent. Res. *51,* 882 (1972).
76 R.J. Gibbons, in: Streptococcus mutans and Dental Caries, p.15. Department of Health, Education and Welfare Publ. No. (NIH) 74-286, Bethesda 1973.
77 B. Guggenheim: Caries Res. *2,* 147 (1968).
78 M.L. Freedman and J.M. Tanzer: Am. Soc. Microbiol. Abstract, April (1972).
79 J.M. Tanzer, in: Streptococcus mutans and Dental Caries, p.25. Department of Health, Education and Welfare Publ. No. (NIH) 74-286, Bethesda 1973.
80 H.V. Jordan: Ann. N.Y. Acad. Sci. *131,* 905 (1965).
81 D.B. Drucker and T.H. Melville: Arch. Oral Biol. *13,* 565 (1968).

82 J.M. Tanzer: Caries Res. *3,* 167 (1969).
83 J.M. Tanzer: J. Dent. Res. *51,* 415 (1972).
84 R.J. Gibbons: Caries Res. *6,* 122 (1972).
85 M.M. McCabe, E.E. Smith and R.A. Cowman: Arch. Oral Biol. *18,* 525 (1973).
86 J.M. Tanzer, A.T. Brown and M.F. McInerney: J. Bacteriol. *116,* 192 (1973).
87 J.M. Tanzer, A.T. Brown and K.I. Myers: 50th Session Int. Assoc. Dental Res., Abstract No.232 (1972).
88 K.K. Mäkinen and A. Scheinin: Int. Dent. J. *21,* 331 (1971).
89 K.K. Mäkinen and A. Scheinin: Acta Odontol. Scand. *30,* 259 (1972).
90 K.K. Mäkinen and A. Scheinin: Acta Odontol. Scand. *32,* 413 (1974).
91 K.K. Mäkinen and A. Scheinin: Acta Odontol. Scand. *33,* Suppl. 70, 129 (1975).
92 C. Mouton, A. Scheinin and K.K. Mäkinen: Acta Odontol. Scand. *33,* 27 (1975).
93 C. Mouton, A. Scheinin and K.K. Mäkinen: Acta Odontol. Scand. *33,* 33 (1975).
94 C. Mouton, A. Scheinin and K.K. Mäkinen: Acta Odontol. Scand. *33,* 251 (1975).
95 A.T. Brown, in: Streptococcus mutans and Dental Caries, p.33. Department of Health, Education and Welfare Publ. No. (NIH) 74-286, Bethesda 1973.
96 R.J. Fitzgerald and P.H. Keyes: J. Am. Dent. Assoc. *61,* 9 (1960).
97 S. Edwardsson: Arch. Oral Biol. *13,* 637 (1968).
98 S. Edwardsson: Odontol. Revy *21,* 153 (1970).
99 A. Scheinin, K.K. Mäkinen and K. Ylitalo: Acta Odontol. Scand. *32,* 383 (1974).
100 A. Scheinin, K.K. Mäkinen and K. Ylitalo: Acta Odontol. Scand. *33,* Suppl. 70, 67 (1975).
101 A. Scheinin, K.K. Mäkinen, E. Tammisalo and M. Rekola: Acta Odontol. Scand. *33,* 269 (1975).
102 K.K. Mäkinen: Int. Dent. J. *26,* 14 (1976).
103 S. Kaaber: Acta Odontol. Scand. *32,* Suppl. 66 (1974).
104 B.N. Ames and R.G. Martin: Annu. Rev. Biochem. *33,* 235 (1964).
105 N.H. Horowitz and R.L. Metzenberg: Annu. Rev. Biochem. *34,* 527 (1965).
106 H.R. Mahler and E.H. Cordes: Biological Chemistry, chap.18. Harpes and Row, New York 1966.
107 R.F. Steiner: The Chemical Foundations of Molecular Biology, chap.1 and 11. Van Nostrand, Princeton 1965.
108 J.D. Watson: Molecular Biology of the Gene, chap. 7–14. Benjamin, New York 1968.
109 K.K. Mäkinen: J. Dent. Res. *51,* 403 (1972).
110 M.L.E. Knuuttila and K.K. Mäkinen: Caries Res. *9,* 177 (1975).
111 M.L.E. Knuuttila and K.K. Mäkinen: 53rd Session Int. Assoc. Dental Res., Abstract No.L79 (1975).
112 K.K. Mäkinen, A. Ojanotko and H. Vidgrén: J. Dent. Res. *54,* 1239 (1975).
113 K.K. Mäkinen and M. Rekola: J. Dent. Res. *55,* 900 (1976).
114 M. Larmas, K.K. Mäkinen and A. Scheinin: Acta Odontol. Scand. *32,* 423 (1974).
115 F. Gehring, K.K. Mäkinen, M.Larmas and A. Scheinin: Acta Odontol. Scand. *32,* 435 (1974).
116 Th.M. Hassell: Dtsch. Zahnärztl. Z. *26,* 1145 (1971).
117 E. Karle and F. Gehring: 11th Annual Meeting Int. Assoc. Dental Res. (CED), Abstracts of Research Reports 2 (1974).
118 E. Karle and F. Gehring: Dtsch. Zahnärztl. Z. *30,* 356 (1975).
119 F. Gehring: Dtsch. Zahnärztl. Z. *26,* 1162 (1971).
120 H.-J. Gülzov: Dtsch. Zahnärztl. Z. *29,* 772 (1974).
121 H.-J. Gülzov: Int. J. Vitam. Nutr. Res. Suppl. *15.* 348 (1976).
122 K. Stegmeier, E. Dallmeier, H.-J. Bestmann and A. Kröncke: Dtsch. Zahnärztl. Z. *26,* 1129 (1971).
123 A.T. Brown and C.E. Patterson: Arch. Oral Biol. *18,* 127 (1973).
124 A.T. Brown and C.L. Wittenberger: Arch. Oral Biol. *18,* 117 (1973).
125 A.T. Brown and C.E. Patterson: Infect. Immun. *6,* 422 (1972).
126 A. Scheinin and K.K. Mäkinen: Int. Dent. J. *21,* 302 (1971).
127 A. Scheinin and K.K. Mäkinen: Acta Odontol. Scand. *30,* 235 (1972).

128 S.B. Horwitz and N.O. Kaplan: J. Biol. Chem. *239*, 830 (1964).
129 M.S. Fox and R.D. Hotchkiss: Nature *187*, 1002 (1960).
130 M.S. Fox: Proc. Natl. Acad. Sci. USA *48*, 1043 (1962).
131 R. Austrian and H.P. Bernheimer: J. Exp. Med. *110*, 571 (1959).
132 G.T. Mills and E.E.B. Smith: Fed. Proc. *21*, 1089 (1962).
133 K.K. Mäkinen and A. Scheinin: Acta Odontol. Scand. *33*, Suppl. 70, 105 (1975).
134 T.T. Wu, E.C.C. Lin and S. Tanaka: J. Bacteriol. *96*, 447 (1968).
135 H.A. Altermatt, F.J. Simpson and A.C. Neish: Can. J. Biochem. Physiol. *33*, 615 (1955).
136 B. Magasanik, M.S. Brooke and D.Karibian: J. Bacteriol. *66*, 611 (1961).
137 S.A. Lerner, T.T. Wu and E.C.C. Lin: Science *146*, 1313 (1964).
138 R.P. Mortlock, D.D. Fossitt and W.A. Wood: Proc. Natl. Acad. Sci. USA *54*, 572 (1965).
139 E.C.C. Lin: J. Biol. Chem. *236*, 31 (1961).
140 B.K. Bhuyan and F.J. Simpson: Can. J. Microbiol. *8*, 737 (1962).
141 D.D. Fossitt, R.P. Mortlock, R.L. Anderson and W.A. Wood: J. Biol. Chem. *239*, 2110 (1964).
142 D.D. Fossitt and W.A. Wood: Methods Enzymol. *9*, 180 (1966).
143 H.J. Fromm: J. Biol. Chem. *233*, 1049 (1958).
144 H.J. Fromm and J.A. Bietz: Arch. Biochem. Biophys. *115*, 510 (1966).
145 H.J. Fromm and D.R. Nelson: J. Biol. Chem. *237*, 215 (1962).
146 S.B. Hulley, S.B. Jorgensen and E.C.C. Lin: Biochim. Biophys. Acta *67*, 219 (1963).
147 R.P. Mortlock, D.D. Fossitt, D.H. Petering and W.A. Wood: J. Bacteriol. *89*, 129 (1965).
148 R.P. Mortlock and W.A. Wood: J. Bacteriol. *88*, 838 (1964).
149 R.P. Mortlock and W.A. Wood: J. Bacteriol. *88*, 845 (1964).
150 R.C. Nordlie and H.J. Fromm: J. Biol. Chem. *234*, 2523 (1959).
151 W.A. Wood, M.J. McDonough and L.B. Jacobs: J. Biol. Chem. *236*, 2190 (1961).
152 L. Marcus and A.G. Marr: J. Bacteriol. *82*, 224 (1961).
153 C. Chiang and S.G. Knight: Biochim. Biophys. Acta *35*, 454 (1959).
154 J.C. Sowden, in: The Carbohydrates, Chemistry, Biochemistry, Physiology, p.77. Ed. W. Pigman. Academic Press, New York 1962.
155 M. Iwasaki: J. Agr. Chem. Soc. Jpn. *16*, 148 (1940).
156 F. Arnal-Peyrot and J. Adrian: Int. J. Vitam. Nutr. Res. *44*, 543 (1974).
157 H.R. Kaback: Annu. Rev. Biochem. *39*, 561 (1970).
158 H.R. Kaback, in: Current Topics in Membranes and Transport, vol. 1, p.35. Eds. F. Bonner and A. Kleinzeller. Academic Press, New York 1970.
159 R. Coleman: Biochim. Biophys. Acta *300*, 1 (1973).
160 K. Yamanaka: Arch. Biochem. Biophys. *131*, 502 (1969).
161 K. Yamanaka: Biochim. Biophys. Acta *151*, 670 (1968).
162 S.M. Cantor and Q.P. Peniston: J. Am. Chem. Soc. *62*, 2113 (1940).
163 B.L. Scallet, K. Shieh, I. Ehrenthal and L. Slapshak: Stärke (Starch) *26*, 405 (1974).
164 J.L. Webb: Enzyme and Metabolic Inhibitors, vol.I, p.199. Academic Press, New York 1963.
165 D.M. Blow and T.A. Steitz: Annu. Rev. Biochem. *39*, 63 (1970).
166 D.E. Koshland, Jr., in: The Enzymes, vol.I, chap.7. Eds. P.D. Boyer, H. Lardy and K. Myrbäck. Academic Press, New York 1959.
167 D.E. Koshland, Jr.: Proc. Natl. Acad. Sci. USA *44*, 98 (1958).
168 A. Gottschalk: Adv. Carbohydr. Chem. *5*, 49 (1950).
169 B. Helferich: Enzymforsch. *2*, 74 (1933).
170 W.W. Pigman: Adv. Enzymol. *4*, 41 (1944).
171 A. Sols and R.K. Crane: J. Biol. Chem. *210*, 581 (1954).
172 W. Kundig, S. Ghosh and S. Roseman: Proc. Natl. Acad. Sci. USA *52*, 1067 (1964).
173 W. Kundig, F.D. Kundig, B.E. Anderson and S.Roseman: Fed. Proc. *24*, 658 (1965).
174 W. Kundig, F.D. Kundig, B.E. Anderson and S.Roseman: J. Biol. Chem. *241*, 3243 (1966).
175 W. Kundig and S. Roseman: Methods Enzymol. *9*, 396 (1966).
176 S.J. Challacombe, T. Lehner and B. Guggenheim: Nature *238*, 219 (1972).
177 R.C. Williams and R.J. Gibbons: Science *177*, 697 (1972).
178 B. Blomberg, W.R. Geckeler and M.Weigert: Science *177*, 178 (1972).

179 E.A. Kabat: Fed. Proc. *21,* 694 (1962).
180 I.C. Grubb: J. Dent. Res. *24,* 31 (1945).
181 D.E. Shaw: J. Dent. Res. *33,* 717 (1954).
182 T.E. Shockley, C.I. Randels and M.C. Dodd: J. Dent. Res. *35,* 233 (1956).
183 M.C. Crowley, V. Harner, A.S. Bennet and P. Jay: J. Am. Dent. Assoc. *52,* 148 (1956).
184 L.S. Fosdick, H.R. Englander, K.C. Hoerman and R.G. Kesel: J. Am. Dent. Assoc. *55,* 191 (1957).
185 G. Frostell: Sverig. Tandläk.-Förb. Tidn. *57,* 696 (1965).
186 H.R. Mühlemann: Schweiz. Monatsschr. Zahnheilk. *79,* 117 (1969).
187 C.E. Klapper and J.F. Volker: J. Dent. Res. *33,* 666 (1954).
188 C.E. Klapper and J.F. Volker: J. Dent. Res. *34,* 703 (1955).
189 J.H. Shaw and D. Griffiths: J. Dent. Res. *39,* 377 (1960).
190 G. Frostell, P.H. Keyes and R.H. Larson: J. Nutr. *93,* 65 (1967).
191 I.J. Møller and S. Poulsen: Community Dent. Oral Epidemiol. *1,* 58 (1973).
192 J. Ainamo, M. Sjöblom, A. Talari and L. Tiainen: J. Clin. Periodontol, in press.
193 G. Frostell: Acta Odontol. Scand. *22,* 457 (1964).
194 H.-J. Gülzov, in: Vergleichende biochemische Untersuchungen über den Abbau des Sorbit durch Mikroorganismen der Mundhöhle. Hanser, München 1968.
195 H.R. Mühlemann, B. Regolati and T.M. Marthaler: Helv. Odontol. Acta *14,* 48 (1970).
196 E. Karle and W. Büttner: Dtsch. Zahnärztl. Z. *26,* 1097 (1971).
197 E. Grunberg, G. Beskid and M. Brin: Int. J. Vitam. Nutr. Res. *43,* 227 (1973).
198 J. Navia, H. Lopez and J. Fischer: 52nd Session Int. Assoc. Dental Res., Abstract No. 611 (1974).
199 O. Larje and R.H. Larson: Arch. Oral Biol. *15,* 805 (1970).
200 U. Böhmer, H.J. Kater, P.-Th. Schneider, H. Welzbacher and W. Büttner: J. Dent. Res. *50,* 752 (1971).
201 H. Gräf: Schweiz. Monatsschr. Zahnheilkd. *79,* 146 (1969).
202 W.H. Bowen, J.E. Eastoe and D.J. Cock: Arch. Oral Biol. *11,* 833 (1966).
203 R. Clark, D.I. Hay, C.J. Schram and B.J. Wagg: Br. Dent. J. *111,* 244 (1961).
204 G. Frostell: Nutrition and Caries Prevention, p. 60. Almquist and Wiksell, Stockholm 1965.
205 D.E.R. Cornick and W.H. Bowen: Arch. Oral Biol. *17,* 1637 (1972).
206 K.K. Mäkinen, W.H. Bowen, D. Dalgard and G.F. Fitzgerald: 54th Session Int. Assoc. Dental Res., Abstract No. 253 (1976).
207 Y. Ericsson: Acta Odontol. Scand. *8,* Suppl. 3 (1949).
208 L.S. Fosdick and A.C. Starke: J. Dent. Res. *18,* 417 (1939).
209 P. Grøn: Arch. Oral Biol. *18,* 1385 (1973).
210 J.E. Hills and H.R. Sullivan: Aust. Dent. J. *3,* 101 (1958).
211 H. Berggren: Sven. Tandläk. Tidn. (Swed. Dental J.) *40,* No. 1B (1947).
212 H. Berggren and H. Hedström: Swed. Dental Assoc. Festival Edition, p. 7 (1950). Sveriges Tandläkarförbund, Stockholm 1950.
213 F. Brudevold, P. Grøn and H. McCann, in: Advances in Fluorine Research and Dental Caries Prevention, p. 63. Eds. J.L. Hardwick, H.R. Held and K.-G. König. Pergamon Press, London 1965.
214 T. Koulourides, F. Feagin and W. Pigman: Ann. N.Y. Acad. Sci. *131,* 751 (1965).
215 I.L. Shannon, R.P. Suddick and F.J. Dowd, Jr.: Monographs in Oral Science, vol. 2. Karger, Basel 1974.
216 P. Grøn: Arch. Oral Biol. *18,* 1365 (1973).
217 A.Y. Balekjian, R.W. Longton and H.D. Tow: 50th Session Int. Assoc. Dental Res., Abstract No. 946 (1972).
218 S.A. Leach, P. Critchley, A.B. Kolendo and C.A. Saxton: Caries Res. *1,* 104 (1967).
219 R.E. Wuthier, P. Grøn and I.T. Irwing: Biochem. J. *92,* 205 (1964).
220 K.K. Mäkinen, P. Lönnberg and A. Scheinin: Acta Odontol. Scand. *33,* Suppl. 70, 277 (1975).
221 R.G. Bibby: Caries Res. *5,* 305 (1971).
222 P. Grøn: Arch. Oral Biol. *18,* 1379 (1973).

223 K.K. Mäkinen, J. Tenovuo and A. Scheinin: Acta Odontol. Scand. *33,* Suppl. 70, 247 (1975).
224 K.K. Mäkinen and P.-L. Mäkinen: 24th Congress of ORCA, Megeve 1977.
225 T. Koulourides, R. Bodden, S. Keller, L. Manson-Hing, J. Lastra and T. Housch: Caries Res. *10,* 427 (1976).
226 D.A. Johnson and L.M. Sreebny: Arch. Oral Biol. *18,* 1555 (1973).
227 M.L.E. Knuuttila and K.K. Mäkinen, to be published.
228 J.M. Tanzer, M.I. Krichevsky and P.H. Keyes: J. Gen. Microbiol. *55,* 351 (1969).
229 C.J. Palenik and C.H. Miller: J. Dent. Res. *54,* 186 (1975).
230 K.K. Mäkinen and A. Scheinin: Acta Odontol. Scand. *33,* Suppl. 70, 317 (1975).
231 I. Kleinberg: Adv. Oral Biol. *4,* 44 (1970).
232 J.L. McDonald, Jr., and G.K. Stookey: J. Dent. Res. *54,* 187 (1975).
233 K.K. Mäkinen, I. Läikkö, A. Scheinin and K.U. Paunio: Acta Odontol. Scand. *33,* Suppl. 70, 297 (1975).
234 K.K. Mäkinen and T. Hyyppä: Arch. Oral Biol. *20,* 509 (1975).
235 M. Morrison and W.F. Steele, in: Biology of the Mouth, p. 89. Ed. P. Person. American Association for Advancement of Science, Washington 1968.
236 H. Hoogendoorn: Academic Dissertation. Technische Hogeschool Delft. Mouton, Den Haag 1974.
237 S.M. Siegel, in: Biology of the Mouth, p. 111. Ed. P. Person. American Association for Advancement of Science, Washington 1968.
238 D. Afonsky, in: Saliva and its Relation to Oral Health, p. 154. University of Alabama Press, Alabama 1961.
239 B.T. Squares: J. Physiol., Lond. *119,* 153 (1953).
240 Z. Dische, C. Pallavicini, H. Kawasaki, N. Smirnow, L. Cizek and S. Chien: Arch. Biochem. Biophys. *97,* 459 (1962).
241 Z. Dische, C.M. Burgher, A. Danilchenko and C. Rothschild: Arch. Biochem. Biophys. *135,* 1 (1969).
242 M.M. Ferguson: 53rd Session Int. Assoc. Dental Res., Abstract No. L1 (1975).
243 D.C. Kroeger and W. Krivory, in: Hypotensive Peptides, p. 289. Eds. E.G. Erdös, N. Back and F. Sicuteri. Springer, New York 1966.
244 K.K. Mäkinen, W.H. Bowen, D. Dalgard and G.F. Fitzgerald: J. Nutr., in press.
245 J.L. Bird, B.J. Baum, K.K. Mäkinen, W.H. Bowen and R.W. Longton: J. Nutr. *107,* 1763 (1977).
246 K.U. Paunio, K.K. Mäkinen and A. Scheinin: Acta Odontol. Scand. *29,* 583 (1971).
247 K.U. Paunio, K.K. Mäkinen and A. Scheinin: Acta Odontol. Scand. *31,* 193 (1973).
248 K.U. Paunio, K.K. Mäkinen and A. Scheinin: Acta Odontol. Scand. *33,* Suppl. 70, 217 (1975).
249 S. Hollmann and O. Touster: J. Am. Chem. Soc. *78,* 3544 (1956).
250 O. Touster, R.M. Hutcheson and V.H. Reynolds: J. Am. Chem. Soc. *76,* 5005 (1954).
251 O. Touster, R.M. Hutcheson and L. Rice: J. Biol. Chem. *215,* 677 (1955).
252 S. Futterman and J.H. Roe: J. Biol. Chem. *215,* 257 (1955).
253 O. Touster, V.H. Reynolds and M. Hutcheson: J. Biol. Chem. *221,* 697 (1956).
254 P.A. Srere, J.R. Cooper, V. Klybas and E. Racker: Arch. Biochem. Biophys. *59,* 535 (1955).
255 B.L. Horecker, J. Hurwitz and P.Z. Smyrniotis: J. Am. Chem. Soc. *78,* 692 (1956).
256 S. Hollmann and O. Touster: J. Biol. Chem. *225,* 87 (1957).
257 O. Touster and D.R.D. Shaw: Physiol. Rev. *42,* 181 (1962).
258 B.L. Horecker: Med. Ernähr. *10,* 66, 95, 127 (1969).
259 S. Hollmann: Hoppe Seylers Z. Physiol. Chem. *317,* 193 (1959).
260 J.J. Burns, J. Kanfer and G. Ashwell: Biochim. Biophys. Acta *34,* 464 (1959).
261 S.M. Nasrallah and F.L. Iber: Am. J. Med. Sci. *25,* 80 (1969).
262 H.E. Ginn, in: Sugars in Nutrition, p. 607. Eds. H.L. Sipple and K.W. McNutt. Academic Press, New York 1974.
263 D.W. Thomas, J.B. Edwards and R.G. Edwards: New Engl. J. Med. *283,* 437 (1970).
264 J.F. Donahoe and R.J. Powers: New Engl. J. Med. *282,* 690 (1970).

265 D.W. Thomas, J.B. Edwards, J.E. Gilligan, J.R. Lawrence and R.G. Edwards: Med. J. Aust. *1,* 1238 (1972).
266 O. Touster, in: Sugars in Nutrition, p.229. Eds. H.L. Sipple and K.W. McNutt. Academic Press, New York 1974.
267 C. C. Arsenis and O. Touster: J. Biol. Chem. *244,* 3895 (1969).
268 R.J. Oshinsky, Y.M. Wang and J. van Eys: Fed. Proc. Abstracts *36,* No. 1183, 419 (1976).
269 I. Kupke and W. Lamprecht: Hoppe Seylers Z. Physiol. Chem. *348,* 17 (1967).
270 D.B. McCormick and O. Touster: J. Biol. Chem. *229,* 451 (1957).
271 J.J. Burns and J. Kanfer: J. Am. Chem. Soc. *79,* 3604 (1957).
272 S.D. Gupta, C.R. Chaudhuri and I.B. Chatterjee: Arch. Biochem. Biophys. *152,* 889 (1972).
273 O. Touster: Fed. Proc. *19,* 977 (1960).
274 H.H. Hiatt, in: Metabolic Basis of Inherited Disease, 3rd ed., p. 119. Eds. J.B. Stanbury, J.B. Wyngaarden and D.S. Fredrickson. McGraw-Hill, New York 1972.
275 S. Hollmann and O. Touster, in: Non-Glycolytic Pathways of Metabolism of Glucose, p. 107. Academic Press, New York 1964.
276 S. Hollmann and O. Touster: Biochim. Biophys. Acta *62,* 338 (1962).
277 T. Asakura, K. Adachi, S. Minakami and H. Yoshikawa: J. Biochem., Tokyo *62,* 184 (1967).
278 Y.M. Wang and J. van Eys: New Engl. J. Med. *282,* 892 (1970).
279 J. van Eys, Y.M. Wang, S.Chan, V.S. Tanphaichitr and S.M. King, in: Sugars in Nutrition, p.613. Eds. H.L. Sipple and K.W. McNutt, Academic Press, New York 1974.
280 M.G. Smith: Biochem. J. *83,* 135 (1962).
281 J.R. Williamson, A. Jacob and C. Refino: J. Biol. Chem. *246,* 7632 (1971).
282 G. Ashwell: Methods Enzymol. *5,* 208 (1962).
283 E.R. Froesch and A. Jacob, in: Sugars in Nutrition, p.241. Eds. H.L. Sipple and K.W. McNutt. Academic Press, New York 1974.
284 U. Keller and E.R. Froesch: Schweiz. Med. Wochenschr. *102,* 1017 (1972).
285 H. Förster: Med. Ernähr. *13,* 7 (1972).
286 H. Förster, H. Hoffmann and I. Hoos: Z. Ernährungswiss. Suppl. *15,* 28 (1973).
287 H. Mehnert, J.D. Summa and H. Förster: Klin. Wochenschr. *42,* 382 (1964).
288 A. Jacob, J.R. Williamson and T. Asakura: J. Biol. Chem. *246,* 7623 (1971).
289 K. Lang: Klin. Wochenschr. *49,* 233 (1971).
290 B.L. Horecker, K. Lang and Y. Takagi (Eds.): Int. Symp. on Metabolism, Physiology, and Clinical Uses of Pentoses and Pentitols. Springer, Berlin 1969.
291 E.R. Froesch, J. Zapf, U. Keller and O. Oelz: Eur. J. Clin. Invest. *2,* 8 (1971).
292 H. Förster, in: Sugars in Nutrition, p.259. Eds. H.L. Sipple and K.W. McNutt. Academic Press, New York 1974.
293 H. Förster, E. Meyer and M. Ziege: Klin. Wochenschr. *48,* 878 (1970).
294 J. Huttunen, K.K. Mäkinen and A. Scheinin: Acta Odontol. Scand. *33,* Suppl. 70, 239 (1975).
295 H. Förster and M. Ziege: Z. Ernährungswiss. *10,* 524 (1971).
296 H. Förster, S. Boecker and M. Ziege: Med. Ernähr. *13,* 193 (1972).
297 R.G. Edwards and J.B. Edwards: Technicon Symp., Frankfurt 1971.
298 W. Schumer: Metabolism *20,* 345 (1971).
299 G. Heraud: Réunions Internationales sur le Sucre, Paris 1973.
300 H.C. Meng, in: Sugars in Nutrition, p.527. Eds. H.L. Sipple and K.W. McNutt. Academic Press, New York 1974.
301 H. Popper and F. Schaffner, in: Die Leber, p.574. Thieme, Stuttgart 1961.
302 K.K. Mäkinen and A. Scheinin: Acta Odontol. Scand. *33,* Suppl. 70, 265 (1975).
303 K.K. Mäkinen, H. Mielityinen and A. Scheinin: Acta Odontol. Scand. *33,* Suppl. 70, 293 (1975).
304 H.K. Prins and J.A. Loos, in: Biochemical Methods in Red Cell Genetics, p. 115. Ed. J.J. Yunis. Academic Press, New York 1969.
305 S.K. Srivastava: Exp. Eye Res. *11,* 294 (1971).
306 H.E. Sauberlich, J.H. Judd, G.E. Nicholas, H.P. Broquist and W.J. Darby: Am. J. Clin. Nutr. *25,* 756 (1972).

307 P.A. Marks and J. Banks: Ann. N.Y. Acad. Sci. *128*, 198 (1965).
308 D.F. Keller: CRC Crit. Rev. Clin. Lab. Sci. *1*, 247 (1970).
309 A. Yoshida, E. Beutler and A.G. Motulsky: Bull. WHO *45*, 243 (1971).
310 K.A. Geser, H. Förster, H. Pröls and H. Mehnert: Klin. Wochenschr. *45*, 851 (1967).
311 T. Asakura, H. Ninomiya, S. Minakamis and H. Yoshikawa, in: Int. Symp. on Metabolism, Physiology, and Clinical Use of Pentoses and Pentitols, p.158. Eds. B.L. Horecker, K. Lang and Y. Takagi. Springer, Berlin 1969.
312 T. Asakura, K. Adachi and H. Yoshikawa: J. Biochem. *67*, 731 (1970).
313 H. Yoshikawa, in: Metabolism and Membrane Permeability of Erythrocytes and Thrombocytes, p.37. Eds. E. Deutsch, E. Gerlach and K. Moser. Thieme, Stuttgart 1970.
314 Y.M. Wang and J. van Eys: New Engl. J. Med. *282*, 892 (1970).
315 T. Asakura, K. Adachi, S. Minakami and H. Yoshikawa: Z. Gesamte Exp. Med. *145*, 266 (1968).
316 K.H. Bässler and W.V. Reimold: Klin. Wochenschr. *43*, 169 (1965).
317 P. Vuopio, M. Härkönen, R. Johnsson and M. Nuutinen: Ann. Clin. Res. *5*, 168 (1973).
318 M. Härkönen and P. Vuopio: Ann. Clin. Res. *6*, 187 (1974).
319 P. Vuopio, M. Härkönen, T. Helske and H. Näveri: Scand. J. Haematol. *15*, 145 (1975).
320 J.H. Kinoshita: Invest. Ophthalmol. *4*, 786 (1965).
321 J.H. Kinoshita: Invest. Ophthalmol. *13*, 713 (1974).
322 H.R. Koch, O. Hockwin and F.I. Weigelin: J. Med. *8*, 1562 (1972).
323 E. Cotlier and B. Becker: Invest. Ophthalmol. *4*, 806 (1965).
324 J.W. Bettman, W.E. Fung and P.O. Noyes: Invest. Ophthalmol. *3*, 678 (1964).
325 J.H. Kinoshita, L.O. Merola and B. Tung: Exp. Eye Res. *7*, 80 (1968).
326 T. Kuwabara and J.H. Kinoshita: Invest. Ophthalmol. *8*, 133 (1969).
327 J.H. Kinoshita, L.O. Merola, K. Satok and E. Dimak: Nature *194*, 1085 (1962).
328 J.H. Kinoshita, L.O. Merola and E. Dimak: Biochim. Biophys. Acta *62*, 176 (1962).
329 J.H. Kinoshita, L.O. Merola and E. Dimak: Exp. Eye Res. *1*, 405 (1962).
330 J.H. Kinoshita, S. Futterman, K. Satok and L.O. Merola: Biochim. Biophys. Acta *74*, 340 (1963).
331 J.H. Kinoshita and L.O. Merola: Invest. Ophthalmol. *3*, 577 (1964).
332 H. Obazawa, L.O. Merola and J.H. Kinoshita: Invest. Ophthalmol. *13*, 204 (1974).
333 L.T. Chylack and J.H. Kinoshita: Invest. Ophthalmol. *8*, 401 (1969).
334 J.H. Kinoshita, W.G. Barber and L.O. Merola: Invest. Ophthalmol. *8*, 133 (1969).
335 R. van Heyningen: Biochem. J. *73*, 197 (1959).
336 J.E. Harris and L.B. Gehrsitz: Am. J. Ophthalmol. *32*, part II, 167 (1949).
337 A. Pirie and R. van Heyningen: Exp. Eye Res. *3*, 124 (1964).
338 D.J. Heaf and D.J. Galton: Clin. Chim. Acta *63*, 41 (1975).
339 R. Gitzelmann, H.C. Curtius and I. Schneller: Exp. Eye Res. *6*, 1 (1967).
340 A. Pirie: Invest. Ophthalmol. *4*, 629 (1965).
341 V.H. Müller and B. Weber: Klin. Monatsbl. Augenheilkd. *158*, 627 (1971).
342 V.R. Marquardt and H. Kirschbaum: Klin. Monatsbl. Augenheilkd. *158*, 769 (1971).
343 R.G. Hansen, R.K. Bretthauser. J. Mayes and J.H. Nordin: Proc. Soc. Exp. Biol. Med. *115*, 560 (1964).
344 G.M. Guest: JAMA *168*, 2015 (1958).
345 L.E. Rosenberg, A.N. Weinberg and S. Segal: Biochim. Biophys. Acta *48*, 484 (1961).
346 W.W. Wells, T.A. Pittman, H.S. Wells and T.S. Egan: J. Biol. Chem. *240*, 1002 (1965).
347 D.W. Thomas, J.E. Gilligan, J.B. Edwards and R.G. Edwards: Med. J. Aust. *1*, 1246 (1972).
348 D.W. Thomas, J.B. Edwards and R.G. Edwards, in: Sugars in Nutrition, p.567. Eds. H.L. Sipple and K.W. McNutt. Academic Press, New York 1974.
349 G.W. Evans, G. Phillips, T.M. Mukherjee, M.R. Snow, J.R. Lawrence and D.W. Thomas: J. Clin. Pathol. *26*, 32 (1972).
350 H.U. Zollinger, in: Spezielle pathologische Anatomie, p.267. Eds. W. Doerr and E. Uehlinger. Springer, Berlin 1966.
351 A. Wretlind: Nutr. Metab. *18*, Suppl. 1, 242 (1975).
352 H. Förster: Dtsch. Med. Wochenschr. *98*, 839 (1973).

353 M. Brin and O.N. Miller, in: Sugars in Nutrition, p.591. Eds. H.L. Sipple and K.W. McNutt. Academic Press, New York 1974.
354 B. Bennett and C. Roseblum: J. Lab. Invest. *10,* 947 (1961).
355 D.W. Thomas, B. Hannett, A. Chalmers, A.M. Rofe, J.B. Edwards and R.G. Edwards: Int. J. Vitam. Nutr. Res. Suppl. *15,* 181 (1976).
356 I.I. Sheleketina: Vopr. Pitan. *28,* No. 6, 44 (1969).
357 K.H. Bässler and K. Schultis, in: Total Parenteral Nutrition, p.65. Ed. H. Ghadimi. Wiley & Sons, New York 1975.
358 G. Beneke and K. Paulini, in: Die Bausteine der parenteralen Ernährung. Eds. H. Beisbarth, R. Horatz and P. Rittmeyer. Enke, Stuttgart 1973.
359 H.U. Zollinger, in: Spezielle pathologische Anatomie, p.270. Eds. W. Doerr and E. Uehlinger. Springer, Berlin 1966.
360 H.J. Pesch, F.D. Krampf, H. Weiland and H. Baudisch, in: Verdauung und Stoffwechsel. Aktionen und Interaktionen. Verhandlungsband Dtsch. Gesellsch. Verdauungs- und Stoffwechselkrankheiten. Witzstrock, Baden-Baden 1974.
361 K. Lang: Int. Z. Vitaminforsch. *34,* 117 (1964).
362 D. Hötzel: Dtsch. Zahnärztl. Z. *26,* 1035 (1971).
363 H. Mehnert: Med. Ernähr. *11,* 77 (1970).
364 H. Mehnert, H. Förster, C.A. Geser, M. Haslbeck and K.H. Dehmel, in: Parenteral Nutrition, p.112. Eds. H.C. Meng and D.H. Law. Thomas, Springfield, Ill., 1970.
365 H. Förster: Z. Ernährungswiss. *11,* 227 (1972).
366 K.H. Bässler, in: Parenteral Nutrition, p.96. Eds. H.C. Meng and D.H. Law. Thomas, Springfield, Ill., 1970.
367 Ch. Gärtner: Dtsch. Med. Wochenschr. *95,* 1327 (1970).
368 H. Birnesser, H. Reinauer and S. Hollmann: Diabetologia *9,* 30 (1973).
369 G. Berg, F. Matzkies, H. Heid and W. Fekl: Z. Ernährungswiss. *14,* 163 (1975).
370 F. Matzkies: Z. Ernährungswiss. *14,* 184 (1975).
371 E.R. Froesch: Nutr. Metab. *18,* Suppl. 1, 30 (1975).
372 G. Schlierf, S. Edlich and B. Schnellenberg: Nutr. Metab. *18,* Suppl. 1, 93 (1975).
373 H. Förster and H. Hoffmann: Infusionsther. Klin. Ernähr. *1,* 265 (1974).
374 M. Haslbeck: Infusionsther. Klin. Ernähr. *1,* 569 (1974).
375 H.F. Woods and H.A. Krebs: Biochem. J. *134,* 437 (1973).
376 H.F. Woods: Nutr. Metab. *18,* Suppl. 1, 65 (1975).
377 H. Brinkrolf and K.H. Bässler: Z. Ernährungswiss. *11,* 167 (1972).
378 J.Ch. Bode: Infusionsther. Klin. Ernähr. *1,* 558 (1974).
379 S. Shima, M. Mitsunaga and T. Nakao: Endocrinol. Jpn. *17,* 283 (1970).
380 H. Förster: Z. Ernährungswiss. Suppl. *11,* 49 (1971).
381 G. Berg, H. Bickel and F. Matzkies: Dtsch. Med. Wochenschr. *98,* 602 (1973).
382 J. van Eys and Y.M. Wang: New Engl. J. Med. *286,* 1162 (1972).
383 H. Förster: New Engl. J. Med. *286,* 790 (1972).
384 I.M. Spitz, A.H. Rubenstein, I. Bersohn and K.H. Bässler: Metabolism *19,* 24 (1970).
385 T. Kuzuya, Y. Kanazawa and K. Kosaka: Metabolism *15,* 1149 (1966).
386 K. Schultis, W. Diedrichson and O. Hahn: Med. Ernähr. *11,* 59 (1970).
387 E. Kuhfahl: Acta Biol. Med. Ger. *21,* 711 (1968).
388 F. Müller, E. Strack, E. Kuhfahl and D. Dettmer: Z. Gesamte Exp. Med. *142,* 338 (1967).
389 S. Yamagata, Y. Goto, A. Ohneda, M. Anzai, S. Kawashima, M. Chiba, Y. Maruhama and Y. Yamauchi: Lancet *II,* 918 (1965).
390 K. Opitz: Arch. Pharmak. Exp. Path. *255,* 192 (1966).
391 H. Förster, M. Haslbeck and H. Mehnert: Infusionsther. Klin. Ernähr. *1,* 199 (1974).
392 C.A. Geser: Infusionsther. Klin. Ernähr. *1,* 215 (1974).
393 H.T. Randall, in: Parenteral Nutrition, p.13. Eds. H.C. Meng and D.H. Law. Thomas, Springfield, Ill., 1970.
394 H.-U. Heuckenkamp and N. Zöllner: Infusionsther. Klin. Ernähr. *1,* 565 (1974).
395 B.L. Horecker, in: Int. Symp. on Metabolism, Physiology, and Clinical Use of Pentoses and Pentitols, p.5. Eds. B.L. Horecker, K. Lang and Y. Takagi. Springer, Berlin 1969.

396 T. Asano, M.D. Levitt and F.C. Goetz: Diabetes *21*, Suppl. 1, 350 (1972).
397 J.E.F. Riseman, S.Koretsky and G.E. Altman: Am. J. Cardiol. *15*, 220 (1965).
398 D. Stiller and E. Hempel: Acta Histochem. (Jena) *36*, 404 (1970).
399 D. Stiller and J. Gorski: Histochemie *5*, 407 (1965).
400 P. Faulkner: Biochem. J. *68*, 374 (1958).
401 M.M. Miller and H.B. Lewis: J. Biol. Chem. *98*, 133 (1932).
402 H.H. Hiatt: J. Biol. Chem. *224*, 851 (1957).
403 J.B. Wyngaarden, S. Segal and J.B. Foley: J. Clin. Invest. *36*, 1395 (1957).
404 N. Okoti: Jpn. Med. J. *2*, 23 (1949).
405 H.R. Mühlemann and J. de Boever, in: Dental Plaque, p.179. Ed. W.D. McHugh. Livingstone Press, Edinburgh 1970.
406 R.W. Winters, P.R. Scaglione, G.G. Nahas and M. Verosky: J. Clin. Invest. *43*, 647 (1964).
407 R. Tudisco: Boll. Soc. Ital. Biol. Sper. *36*, 1610 (1960).
408 Y. Kumahara, D.S. Feingold, I.M. Freeberg and H.H. Hiatt: J. Clin. Endocrinol. Metab. *21*, 887 (1961).
409 R.K. Haydon: Biochim. Biophys. Acta *46*, 598 (1961).
410 C.H. Mellinghoff: Klin. Wochenschr. *39*, 447 (1961).
411 W. Kieckebuch, W. Griem and K. Lang: Klin. Wochenschr. *39*, 447 (1961).
412 K.H. Bässler, W. Prellwitz, V. Unbehaun and K. Lang: Klin. Wochenschr. *40*, 791 (1962).
413 W. Griem and K. Lang: Klin. Wochenschr. *40*, 801 (1962).
414 K.H. Bässler and G. Dreiss: Klin. Wochenschr. *41*, 593 (1963).
415 K.H. Bässler and D. Heesen: Klin. Wochenschr. *41*, 595 (1963).
416 G. Czok and K. Lang: Klin. Wochenschr. *41*, 241 (1963).
417 V.G. Foglia, R.Yabo, J.P. Bernaldez and L.M. Aguirre: Folia Endocrinol., Roma *16*, 240 (1963).
418 W. Prellwitz and K.H. Bässler: Klin. Wochenschr. *41*, 196 (1963).
419 H. Mehnert, J.D. Summa and H.Förster: Klin. Wochenschr. *42*, 382 (1964).
420 B. Schmidt, M. Fingerhut and K. Lang: Klin. Wochenschr. *42*, 1073 (1964).
421 F. Manenti and L. Della Casa: Boll. Soc. Med.-Chir. Modena *65*, 1309 (1965).
422 W. Montague, S.L. Howell and K.W. Taylor: Nature *215*, 1088 (1967).
423 K.H. Bässler and W. Prellwitz: Klin. Wochenschr. *42*, 94 (1964).
424 G. Müller: Klin. Wochenschr. *42*, 454 (1964).
425 Y. Akazawa, T. Iehara, H. Mori, B. Hashimoto and T. Tanaka: Iryo Medical Treatment. Med. J. Natl. Hospitals Sanatoriums Jpn., Tokyo *21*, 891 (1967).
426 K.H. Bässler, H. Holzmann, W. Prellwitz and J. Stechert: Klin. Wochenschr. *43*, 27 (1965).
427 Y. Hirata, M. Fujisawa, H. Sato, T. Asano and S. Katsuki: Biochem. Biophys. Res. Commun. *24*, 471 (1966).
428 K. Lang, R. Frey and M. Halmágyi: Münch. Med. Wochenschr. *108*, 2163 (1966).
429 K.H. Bässler, G. Stein and W. Belzer: Biochem. Z. *346*, 171 (1966).
430 K.H. Bässler, W. Toussaint and G. Stein: Klin. Wochenschr. *44*, 212 (1966).
431 W. Toussaint, K. Roggenkamp and K.H. Bässler: Z. Kinderheilkd. *98*, 146 (1967).
432 K. Yoshikawa: Masui. Jpn. J. Anesthesiol. *16*, 311 (1967).
433 V.R. Klyachko, A.A. Perelygina and I.V. Domareva: Sov. Med. *31*, 78 (1968).
434 L.F. Podryadkov: Vopr. Pitan. *27*, 19 (1968).
435 Y. Hirata, M. Fujisawa, H. Sato, T. Asano and S. Katsuki: Z. Gesamte Exp. Med. *145*, 111 (1968).
436 W. Montague and K.W. Taylor: Biochem. J. *109*, 333 (1968).
437 S. Paschmi and K. Opitz: Z. Gesamte Exp. Med. *148*, 22 (1968).
438 G. Stein and K.H. Bässler: Z. Gesamte Exp. Med. *147*, 197 (1968).
439 K. Schultis and C.A. Geser: Anaesth. Resusc. Intensive Ther. *31*, 30 (1968).
440 K. Schultis and C.A. Geser, in: Parenteral Nutrition, p.139. Eds. H.C. Meng and D.H. Law. Thomas, Springfield, Ill., 1970.
441 W. Toussaint: Anaesth. Resusc. Intensive Ther. *31*, 38 (1968).
442 C.A. Geser and H.Mehnert: Anaesth. Resusc. Intensive Ther. *31*, 7 (1968).
443 K. Aono: Anaesth. Resusc. Intensive Ther. *31*, 52 (1968).

444 E. Pitkänen and K. Sahlström: Ann. Med. Exp. Fenn. *47,* 143 (1968).
445 M. Halmágyi and H.H. Israng: Anaesth. Resusc. Intensive Ther. *31,* 25 (1968).
446 E. Pitkänen and A. Pitkänen: Ann. Med. Exp. Fenn. *42,* 113 (1964).
447 M. Fujisawa: Igaku Kenkyu, Acta Med., Fukuoka *38,* 105 (1968).
448 M. Kondo, M. Naito and K. Sugiura: Hinyokika Kiyo (Acta Urol. Jpn.) *14,* 215 (1968).
449 H. Iwasa and M. Takahashi: Gan no Rinsho (Jpn. J. Cancer Clin.) *14,* 561 (1968).
450 L.S. Marimjan, H.G. Gusarenko, T.C. Lanshina, J.I. Popova and V.D. Tkatsenko: Vrach. Delo *11,* 148 (1968).
451 H. Ishii, H. Takahashi, H. Mamori and S. Murai: Keio J. Med. *18,* 109 (1969).
452 D.A. Coats: Z. Ernährungswiss. *9,* 401 (1969).
453 L.E. Blockus, J.F. Donahoe, T.A. Crowley, J.W. Keating and M.S. Weinberg: Annual Meeting of the Society of Toxicology, Williamsburg, Virginia, March 9–13, 1969.
454 T. Kuzuya and Y. Kanazawa: Diabetologia *5,* 248 (1969).
455 U.C. Dubach, E. Feiner and I. Forgó: Schweiz. Med. Wochenschr. *99,* 190 (1969).
456 Y. Kinami and I. Kitagawa: Shujutsu (Operation) *23,* 1487 (1969).
457 T. Kuzuya, Y. Kanazawa and K. Kosaka: Endocrinology *84,* 200 (1969).
458 Y. Hirata, M. Fujisawa and T. Ogushi, in: Int. Symp. on Metabolism, Physiology, and Clinical Use of Pentoses and Pentitols, p. 226. Eds. B.L. Horecker, K. Lang and Y. Takagi. Springer, Berlin 1969.
459 N. Hosoya and T. Machiya, in: Int. Symp. on Metabolism, Physiology, and Clinical Use of Pentoses and Pentitols, p. 248. Eds. B.L. Horecker, K. Lang and Y. Takagi. Springer, Berlin 1969.
460 G. Erdmann, in: Int. Symp. on Metabolism, Physiology, and Clinical Use of Pentoses and Pentitols, p. 342. Eds. B.L. Horecker, K. and Y. Takagi. Springer, Berlin 1969.
461 S. Yamagata, Y. Goto, A. Ohneda, M. Anzai, S. Kawashima, J. Kikuchi, M. Chiba, Y. Maruhama, Y. Yamauchi and T. Toyota, in: Int. Symp. on Metabolism, Physiology, and Clinical Use of Pentoses and Pentitols, p. 316. Eds. B.L. Horecker, K. Lang and Y. Takagi. Springer, Berlin 1969.
462 H. Mehnert, H. Förster and K.-H. Dehmel, in: Int. Symp. on Metabolism, Physiology, and Clinical Use of Pentoses and Pentitols, p. 293. Eds. B.L. Horecker, K. Lang and Y. Takagi. Springer, Berlin 1969.
463 M. Uesuka, T. Tosen, M. Ohba, T. Akamatsu, N. Narahara and S. Kodaira, in: Int. Symp. on Metabolism, Physiology, and Clinical Use of Pentoses and Pentitols, p. 388. Eds. B.L. Horecker, K. Lang and Y. Takagi. Springer, Berlin 1969.
464 H. Ishii and K. Sambe, in: Int. Symp. on Metabolism, Physiology, and Clinical Use of Pentoses and Pentitols, p. 309. Eds. B.L. Horecker, K. Lang and Y. Takagi. Springer, Berlin 1969.
465 K. Schultis and C.A. Geser, in: Int. Symp. on Metabolism, Physiology, and Clinical Use of Pentoses and Pentitols, p. 378. Eds. B.L. Horecker, K. Lang and Y. Takagi. Springer, Berlin 1969.
466 K. Opitz, in: Int. Symp. on Metabolism, Physiology, and Clinical Use of Pentoses and Pentitols, p. 238. Eds. B.L. Horecker, K. Lang and Y. Takagi. Springer, Berlin 1969.
467 K.-H. Dehmel, H. Förster and H. Mehnert, in: Int. Symp. on Metabolism, Physiology, and Clinical Use of Pentoses and Pentitols, p. 177. Eds. B.L. Horecker, K. Lang and Y. Takagi. Springer, Berlin 1969.
468 Y. Ohta, T. Takano and K. Kosaka, in: Int. Symp. on Metabolism, Physiology, and Clinical Use of Pentoses and Pentitols, p. 275. Eds. B.L. Horecker, K. Lang and Y. Takagi. Springer, Berlin 1969.
469 M.C. Wang and H.C. Meng: Fed. Proc. *28,* 625 (1969).
470 A.E. Lambert, A. Junod, W. Stauffacher, B. Jeanrenaud and A.E. Renold: Biochim. Biophys. Acta *184,* 529 (1969).
471 N. Hosoya and N. Iitoyo: J. Jpn. Soc. Food Nutr. *22,* 17 (1969).
472 H.P. Wolf, W. Queisser and K. Beck: Klin. Wochenschr. *47,* 1084 (1969).
473 G. Berg, D. Leis and E. Ohnhaus: Med. Ernähr. *10,* 28 (1969).
474 I. Spitz and A. Rubenstein: Med. Ernähr. *11,* 84 (1970).

475 R. Landgraf and F.M. Matschinsky: Fed. Proc. *29* (2), 314 (1970).
476 F.M. Matschinsky and J.E. Ellerman: J. Biol. Chem. *243,* 2730 (1968).
477 I.M. Spitz, I. Bersohn, A.H. Rubenstein and M. van As: Am. J. Med. Sci. *260,* 224 (1970).
478 C.A. Geser, K. Schultis and W. Diedrichson: Med. Ernähr. *11,* 82 (1970).
479 R.L. Swarm and R. Banziger, unpublished, quoted by M. Brin and O.N. Miller, in: Sugars in Nutrition, p.591. Eds. H.L. Sipple and K.W. McNutt. Academic Press, New York 1974.
480 R. Banziger, unpublished, quoted by M. Brin and O.N. Miller, in: Sugars in Nutrition, p.591. Eds. H.L. Sipple and K.W. McNutt. Academic Press, New York 1974.
481 R.B. Wilson and J.M. Martin: Diabetes *19,* 17 (1970).
482 E.M. Bayliss, F. Greenwood, V. James, in: Growth Hormone, p.89. Eds. A. Pecile and E.E. Muller. Excerpta Medica Foundation, No.158, Amsterdam 1968.
483 G. Strauch, P. Pandos and H. Bricaire: J. Clin. Endocrinol. Metab. *32,* 582 (1971).
484 A. Jacob, T. Asakura and J.R. Williamson: Diabetes *19,* Suppl. 1, 357 (1970).
485 F.M. Matschinsky, J. Ellerman, J. Kotler-Brajtburg, J. Krzanowski, R. Fertel and R. Landgraf: Diabetes *19,* Suppl. 1, 365 (1970).
486 W. Grabner, G. Berg, D.Bergner, F. Matzkies and H.-U. Maeder: Med. Ernähr. *12,* 126 (1971).
487 Y.M. Wang, J.H. Patterson and J. van Eys: J. Clin. Invest. *50,* 1421 (1971).
488 R.C. Turner, B. Schneeloch and J.D.N. Nabarro: J. Clin. Endocrinol. Metab. *33,* 301 (1971).
489 W. Pool, unpublished, quoted by M. Brin and O.N. Miller, in: Sugars in Nutrition p.591. Eds. H.L. Sipple and K.W. McNutt. Academic Press, New York 1974.
490 M. Mosinger, unpublished, quoted by M. Brin and O.N. Miller, in: Sugars in Nutrition, p.591. Eds. H.L. Sipple and K.W. McNutt. Academic Press, New York 1974.
491 J.R. Williamson, A. Jakob and R. Scholz: Metabolism *20,* 13 (1971).
492 U. Keller and E.R. Froesch: Diabetologia *7,* 349 (1971).
493 T. Kuzuya, Y. Kanazawa, M. Hayashi, M. Kikuchi and T. Ide: Endocrinol. Jpn. *18,* 309 (1971).
494 F. Amador and A. Eisenstein, unpublished, quoted by M. Brin and O.N. Miller, in: Sugars in Nutrition, p.591. Eds. H.L. Sipple and K.W. McNutt. Academic Press, New York 1974.
495 W. Humke and K.-J. Wiessmann: Arch. Gynäkol. *210,* 97 (1971).
496 P.-U. Heuckenkamp and N. Zöllner: Klin. Wochenschr. *50,* 1063 (1972).
497 H. Willgerodt, K. Beyreiss and H. Theile: Acta Biol. Med. Ger. *28,* 651 (1972).
498 H.W. Staub and R. Thiessen: Fed. Proc. *31,* 291 (1972).
499 D.P. Mertz, V. Kaiser, M. Klöpfer-Zaar and H. Beisbarth: Klin. Wochenschr. *50,* 1107 (1972).
500 D.P. Mertz, V. Kaiser, M. Klöpfer-Zaar and H. Beisbarth: Klin. Wochenschr. *50,* 1097 (1972).
501 Y.M. Wang, S.M. King, J.H. Patterson and J. van Eys: Fed. Proc., Abstract No.2873 (1972).
502 M.H. Jourdan, I. Macdonald and J.R. Henderson: Nutr. Metab. *14,* 92 (1972).
503 Ch. Petrich, H. Reinauer and S. Hollmann: Z. Klin. Chem. Klin. Biochem. *10,* 355 (1972).
504 W. Berger, H. Göschke, J. Moppert and H. Künzli: Horm. Metab. Res. *5,* 4 (1973).
505 T. Asano, M.D. Levitt and F.C. Goetz: Diabetes *22,* 279 (1973).
506 U.C. Dubach, R. Lejeune, I. Forgó and A. Bückert: Dtsch. Med. Wochenschr. *98,* 1960 (1973).
507 W. Büngener: Zentralbl. Allg. Pathol. *117,* 204 (1973).
508 J.B. Blair, D.E. Cook and H.A. Lardy: J. Biol. Chem. *248,* 3601 (1973).
509 H. Förster, I. Hoos and D. Lerche: Diabetologia *9,* 67 (1973).
510 H. Bickel, H. Bünte, D.A. Coats, P. Misch, L. von Rauffer, P. Scranowitz and F. Wopfner: Dtsch. Med. Wochenschr. *98,* 809 (1973).
511 F. Matzkies, W. Grabner, H. Scharrer, H. Bickel and G. Berg: Horm. Metab. Res. *5,* 221 (1973).
512 G. Berg, F. Matzkies and D. Bergner: Klin. Wochenschr. *51,* 1124 (1973).
513 T. Igarashi, M. Kobayashi, T. Shoji, Y. Tanabe, K. Nomura and S. Ohtake: Tohoku J. Exp. Med. *111,* 147 (1973).
514 G. Berg, F. Matzkies and K. Renz: Z. Ernährungswiss. *12,* 159 (1973).

515 M. Wenzel: Z. Ernährungswiss. *12*, 67 (1973).
516 R. Schröder, W. Féaux de Lacroix, U. Franzen, P.J. Klein and W. Müller: Acta Neuropathol., Berl. *27*, 181 (1974).
517 W. Grabner, H.-J. Pesch, F. Matzkies, H. Bickel and G. Berg: Res. Exp. Med. Berl. *162*, 133 (1974).
518 F. Matzkies and G. Berg: Infusionsther. Klin. Ernähr. *1*, 545 (1974).
519 W. Kaiser, K. Stocker and P.-U. Heuckenkamp: Infusionsther. Klin. Ernähr. *1*, 548 (1974).
520 P.-U. Heuckenkamp: Ernähr. Umsch. *21*, 70 (1974).
521 G. Berg, F. Matzkies and H. Bickel: Dtsch. Med. Wochenschr. *99*, 633 (1974).
522 H. Förster and D. Zagel: Dtsch. Med. Wochenschr. *99*, 1300 (1974).
523 H. Wolf, V. Melichar, W. von Berg and J. Kerstan: Infusionsther. Klin. Ernähr. *1*, 479 (1974).
524 C.A. Geser, R. Müller-Hess and J.P. Felber: Infusionsther. Klin. Ernähr. *1*, 483 (1974).
525 K. Schultis, H.J. Gerlach and J.P. Felber: Int. Symp. Growth Hormone, Milan 1971. Exc. Med. Found., Amsterdam 1971.
526 F. Matzkies: Z. Ernährungswiss. *13*, 113 (1974).
527 H. Mehnert, G. Dietze and M. Haslbeck: Nutr. Metab. *18*, Suppl. 1, 171 (1975).
528 P.-U. Heuckenkamp and N. Zöllner: Nutr. Metab. *18*, Suppl. 1, 209 (1975).
529 F. Matzkies, H. Heid, W. Fekl and G. Berg: Z. Ernährungswiss. *14*, 53 (1975).
530 G. Berg, F. Matzkies, H. Heid, M. Fekl and M. Conolly: Z. Ernährungswiss. *14*, 64 (1975).
531 J. Huttunen: Int. J. Vitam. Nutr. Res. Suppl. *15*, 105 (1976).
532 D.W. Thomas, J.B. Edwards and R.G. Edwards: Nutr. Metab. *18*, Suppl. 1, 227 (1975).
533 B.D. Ross, R. Hems and H.A. Krebs: Biochem. J. *102*, 942 (1967).
534 A.T. Maynard: Am. J. Med. Technol. *35*, 412 (1969).
535 H. Göschke, A. Leutenegger and U.F. Gruber: Nutr. Metab. *18*, Suppl. 1, 197 (1975).
536 E. Pitkänen: Clin. Chim. Acta *38*, 221 (1972).
537 Hla-Yee-Yee, M. Mya-Tu and Aung-Than-Batu: Am. J. Clin. Nutr. *26*, 688 (1973).
538 S.M. Partridge: Nature *164*, 443 (1949).
539 K.K. Mäkinen, in: Proceedings Microbial Aspects of Dental Caries. Eds. H.M. Stiles, W.J. Loesche and T.C. O'Brien. Sp. Suppl. Microbiology Abstracts II, 521 (1976).
540 H. Förster, E. Meyer and M. Ziege: Klin. Wochenschr. *48*, 878 (1970).
541 P.-U. Heuckenkamp: Z. Ernährungswiss. *12*, 288 (1973).
542 G. Berg: Z. Ernährungswiss. *12*, 280 (1973).
543 F. Gehring and J. Patz: Dtsch. Zahnärztl. Z. *29*, 1026 (1974).
544 E.J. Karle and F. Gehring: Dtsch. Zahnärztl. Z. *31*, 22 (1976).
545 K.M. Pruitt and M. Adamson: 54th Session Int. Assoc. Dental Res., Abstract No. 123 (1976).
546 M. Adamson and K.M. Pruitt: 54th Session Int. Assoc. Dental Res., Abstract No. 124 (1976).
547 J.A. Kanapka and I. Kleinberg: 54th Session Int. Assoc. Dental Res., Abstract No. 490 (1976).
548 A.J. Narkates, J.M. Navia and D. Bates: 54th Session Int. Assoc. Dental Res., Abstract No. 455 (1976).
549 T. Koulourides, P. Phantumvanit, E.C. Munksgaard, E.C. and T. Housch: J. Oral Pathol. *3*, 185 (1974).
550 T. Koulourides, J. Rundegren and Th. Ericson: 54th Session Int. Assoc. Dental Res., Abstract No. 582 (1976).
551 D.K. Ponton: 54th Session Int. Assoc. Dental Res., Abstract No. 1047 (1976).
552 Y.M. Wang, R.J. Oshinsky, E. Lantin and J. van Eys: Fed. Proc. Abstracts *35*, No. 3970, 920 (1975).
553 H. Mader and H. Reinauer: Z. Ernährungswiss. *15*, 54 (1976).
554 R.H. Ylikahri and T. Leino: Eur. J. Clin. Invest., in press.
555 U.C. Dubach: Int. J. Vitam. Nutr. Res. Suppl. *15*, 81 (1976).
556 F. Matzkies and G. Berg: Int. J. Vitam. Nutr. Res. Suppl. *15*, 48 (1976).
557 H. Bickel: Int. J. Vitam. Nutr. Res. Suppl. *15*, 131 (1976).
558 H. Willgerodt and K. Beyreiss: Int. J. Vitam. Nutr. Res. Suppl. *15*, 161 (1976).
559 K. Paulini: Int. J. Vitam. Nutr. Res. Suppl. *15*, 204 (1976).

560 H.-J. Pesch, F.-D. Krampf, H. Menzel, H. Weiland, U.-W. Eidam, H. Prestele and H. Heid: Int. J. Vitam. Nutr. Res. Suppl. *15*, 193 (1976).

561 R.W.E. Watts, S. Hauschildt, R.A. Chalmers and A.M. Lawson: Int. J. Vitam. Nutr. Res. Suppl. *15*, 216 (1976).

562 R. Dölp and K. Paulini: Int. J. Vitam. Nutr. Res. Suppl. *15*, 228 (1976).

563 F.W. Ahnefeld, M. Halmágyi and P. Milewski: Int. J. Vitam. Nutr. Res. Suppl. *15*, 242 (1976).

564 E.R. Froesch: Int. J. Vitam. Nutr. Res. Suppl. *15*, 239 (1976).

565 H. Göschke, A. Leutenegger and M. Allgöwer: Int. J. Vitam. Nutr. Res. Suppl. *15*, 277 (1976).

566 F. Matzkies and G. Berg: Int. J. Vitam. Nutr. Res. Suppl. *15*, 153 (1976).

567 K. Brand and K.-H. Quadflieg: Int. J. Vitam. Nutr. Res. Suppl. *15*, 44 (1976).

568 E. Söderling, M. Rekola, K.K. Mäkinen and A. Scheinin: Acta Odontol. Scand. *33*, Suppl. 70, 337 (1975).

569 J. Tenovuo and M.L.E. Knuuttila: J. Dent. Res., in press.

570 M. Larmas, A. Scheinin, F. Gehring and K.K. Mäkinen: Acta Odontol. Scand. *33*, Suppl. 70, 321 (1975).

571 H.F. Woods: Int. J. Vitam. Nutr. Res. Suppl. *15*, 54 (1976).

572 A.E. Renold: Int. J. Vitam. Nutr. Res. Suppl. *15*, 52 (1976).

573 H. Förster: Int. J. Vitam. Nutr. Res. Suppl. *15*, 116 (1976).

574 G. Berg, F. Matzkies, H. Bickel and R. Zeilhofer: Z. Ernährungswiss. Suppl. *15*, 47 (1973).

575 L. Heller: Int. J. Vitam. Nutr. Res. Suppl. *15*, 150 (1976).

576 E. Poutiainen, M. Tuori and I. Sirviö: Proc. Nutr. Soc. *35*, 140A (1976).

577 K.H. Bässler: Int. J. Vitam. Nutr. Res. Suppl. *15*, 22 (1976).

578 C.A. Geser: Int. J. Vitam. Nutr. Res. Suppl. *15*, 58 (1976).

579 A. Jakob: Int. J. Vitam. Nutr. Res. Suppl. *15*, 66 (1976).

580 H. Förster: Int. J. Vitam. Nutr. Res. Suppl. *15*, 68 (1976).

581 D.W. Fekl: Int. J. Vitam. Nutr. Res. Suppl. *15*, 76 (1976).

582 K.H. Bässler: Int. J. Vitam. Nutr. Res. Suppl. *15*, 76 (1976).

583 N. Zöllner, P.-U. Heuckenkamp and W. Nechwatal: Klin. Wochenschr. *46*, 1300 (1968).

584 P.A. Poole-Wilson, J. Patrick, G.A. MacGregor and N.F. Jones: Clin. Sci. *43*, 561 (1972).

585 B.J. Stinebaugh, S.A. Bartow, G. Eknoyan, M. Martinez-Maldonado and W.N. Suki: Am. J. Physiol. *220*, 1271 (1971).

586 P.-U. Heuckenkamp: Int. J. Vitam. Nutr. Res. Suppl. *15*, 170 (1976).

587 W. Kaiser, K. Stocker and P.-U. Heuckenkamp, in: Purine Metabolism in Man, 41B, p. 479. Eds. O. Sperling, A. de Vries and J.B. Wyngaarden. Plenum Publ. Corp., New York 1974.

588 E.R. Froesch: Int. J. Vitam. Nutr. Res. Suppl. *15*, 179 (1976).

589 K. Bjondahl and M. Virkki: Nutr. Rep. Int. (1977), in press.

590 T.T. Wu: Biochim. Biophys. Acta *428*, 656 (1976).

591 J.L. Wood, in: Chemistry and Biochemistry of Thiocyanic Acid and Its Derivatives, p. 156. Ed. A.A. Newman. Academic Press, London 1975.

592 J.T. Hindmarsh: Clin. Biochem. *9*, 141 (1976).

593 R. Dölp, E. Grab, E. Knoche and F.W. Ahnefeld: Infusionsther. Klin. Ernähr. *2*, 103 (1975).

594 Y. Seino, T. Taminato, Y. Inoue, Y. Goto, M. Ikeda and H. Imura: J. Clin. Endocrinol. Metab. *42*, 736 (1976).

595 G.E. Demetrakopoulos and H. Amos: Biochem. Biophys. Res. Commun. *72*, 1169 (1976).

596 D.W. McCandless, A.D. Curley and C.E. Cassidy: J. Nutr. *106*, 1144 (1976).

597 M.A. Alizade, K. Brendel and K. Gaede: FEBS Letters *67*, 41 (1976).

598 G. Frostell: Dtsch. Zahnärztl. Z. *32*, Suppl. 1, 71–75 (1977).

599 P. Schneider and H. Mühlemann: Schweiz. Monatsschr. Zahnheilkd. *86*, 150 (1976).

600 F. Gehring and E. Karle: Süsswaren *9*, 330 (1975).

601 E. Karle: Dtsch. Zahnärztl. Z. *32*, Suppl. 1, 89–95 (1977).

602 H. Förster: Dtsch. Zahnärztl. Z. *32*, Suppl. 1, 43–53 (1977).

603 K.K. Mäkinen, P.-L. Mäkinen, E. Söderling and J. Tenovuo, to be published.

604 K.K. Mäkinen, M. Tuori, A. Poutiainen, M. Hämäläinen, to be published.

605 D. Pansu, M.C. Chapuy, M. Milani and C. Bellaton: Calcif. Tissue Res. *21,* Suppl., 45 (1976).
606 L.R. Harper, A.E. Poole and S.I. Wolf: 55th Session Int. Assoc. Dental Res., Abstract No. 78 (1977).
607 K.K. Mäkinen and P.-L. Mäkinen: 55th Session Int. Assoc. Dental Res., Abstract No. 77 (1977).
608 M. Enklewitz and M. Lasker: Am. J. Med. Sci. *186,* 539 (1933).
609 W.E. Knox: Am. J. Hum. Genet. *10,* 385 (1958).
610 L. Sestoft and A. Gammeltoft: Biochem. Pharmacol. *25,* 2619 (1976).
611 T. Noguchi and H.R. Mühlemann: Helv. Odontol. Acta *20,* 45 (1976).
612 E.M. Plüss: J. Clin. Periodontol., in press.
613 I. Kleinberg, J.A. Kanapka and D. Craw: Proceedings Microbial Aspects of Dental Caries. Eds. H.M. Stiles, W.J. Loesche and T.C. O'Brien. Sp. Suppl. Microbiology Abstracts II, 433 (1976).
614 K.K. Mäkinen and K. Virtanen: J. Dent. Res. (1978), in press.
615 M.G. Buonocore: J. Am. Dent. Assoc. *87,* 1000 (1973).
616 W.J. O'Brien: Proceedings Microbial Aspects of Dental Caries. Eds. H.M. Stiles, W.J. Loesche and T.C. O'Brien. Sp. Suppl. Microbiology Abstracts II, 387 (1976).
617 T. Chandra, R. Das and A.G. Datta: Eur. J. Biochem. *72,* 259 (1977).
618 H. Förster, S. Boecker and A. Walther: Fortschr. Med. *95,* 99 (1977).
619 M. Brin, E. Gabriel and L.J. Machlin: Fed. Proc. *36,* Abstr. 1173 (1977).
620 K.C. Hoerman: Fed. Proc. *36,* Abstract 1133 (1977).
621 T. Asano, B.Z. Greenberg, R.V. Wittmers and F.C. Goetz: Endocrinology *100,* 339 (1977).
622 R.M. Green and S.A. Leach: 55th Session Int. Assoc. Dental Res., Abstract No. 207 (1977).
623 H.R. Mühlemann, R. Schmidt, T. Noguchi, T. Imfeld and R.S. Hirsch: Caries Res. *11,* 263 (1977).
624 R.M. Green and S.A. Leach: 55th Session Int. Assoc. Dental Res., Abstract No. 207 (1977).
625 C. Aminoff, E. Vanninen and T. Doty: Int. Symp. on Xylitol, London (1977). In press.
626 F. Voirol: Int. Symp. on Xylitol, London (1977). In press.
627 F. Kracher: Int. Symp. on Xylitol, London (1977). In press.
628 K.H. Bässler: Int. Symp. on Xylitol, London (1977). In press.
629 H. Förster: Int. Symp. on Xylitol, London (1977). In press.
630 K.K. Mäkinen: Int. Symp. on Xylitol, London (1977). In press.
631 F. Gehring: Int. Symp. on Xylitol, London (1977). In press.
632 Bergey's Manual of Determinative Bacteriology. The Williams and Wilkins Company, Baltimore 1974.
633 K.H. Schleifer and W.E. Kloos: Int. J. Syst. Bacteriol. *25,* 50 (1975).
634 K.K. Mäkinen and P.-L. Mäkinen: 24th ORCA Congr., Abstract No. 69 (1977).
635 R. Havenaar, J.H.J. Huis in 't Veld, O. Backer Dirks and J.D. de Stoppelaar: 24th ORCA Congr., Abstract No. 70 (1977).
636 F. Gehring and J.E. Karle: 24th ORCA Congr., Abstract No. 71 (1977).
637 R. Moll and W. Büttner: 24th ORCA Congr., Abstract No. 72 (1977).
638 H.A. Lee and A. Wretlind: Acta Chir. Scand. Suppl. *466,* 6 (1976).
639 L. Heller: Acta Chir. Scand. Suppl. *466,* 10 (1976).
640 W. Hartig, H. Faust and H.-D. Czarnetzki: Acta Chir. Scand. Suppl. *466,* 36 (1976).
641 P.Y. Wang, T.L. Hsu, K.Y. Chien, K.S. Lu and H.C. Meng: Acta Chir. Scand. Suppl. *466,* 46 (1976).
642 F. Lackner, L. Baumgartner und K. Steinenbereithner: Acta Chir. Scand. Suppl. *466,* 48 (1976).
643 K. Schreier and U. Porath: Acta Chir. Scand. Suppl. *466,* 108 (1976).
644 J. Kankare: Anal. Chem. *45,* 2050 (1973).
645 H. Steinberg, in: Organoboron Chemistry. Vol. I, pp. 341, 654, 864. Wiley Interscience, London 1964.
646 S.H. Bok and A.L. Demain: Anal. Biochem. *81,* 18 (1977).
647 F.H. Mattson and G.A. Nolen: J. Nutr. *102,* 1171 (1972).

648 F.H. Mattson and R.A. Volpenhein: J. Nutr. *102,* 1177 (1972).
649 J.F. Donahoe and R.J. Powers: J. Clin. Pharmacol. *14,* 255 (1974).
650 S. Hauschildt, R.A. Chalmers, A.M. Lawson, K. Schultis and R.W.E. Watts: Am. J. Clin. Nutr. *29,* 258 (1976).
651 R.J. Oshinsky, Y.-M. Wang and J. van Eys: J. Nutr. *107,* 792 (1977).
652 Y.-M. Wang, R.J. Oshinsky and J. van Eys: Am. J. Clin. Nutr. *30* (1977). In press.
653 A.F. Leutenegger, H. Göschke, K. Stutz, H. Mannhart, D. Werdenberg, G. Wolf and M. Allgöwer: Am. J. Surg. *133,* 199 (1977).
654 Y.-M. Wang, R.J. Oshinsky, E. Lantin, W. Ukab and J. van Eys: Infusionsther. Klin. Ernähr. (1977). In press.
655 B. Hannett, D.W. Thomas, A.H. Chalmers, A.M. Rofe, J.B. Edwards and R.G. Edwards: J. Nutr. *107,* 458 (1977).